Computer Simulated Experiments for Electric Circuits Using Electronics Workbench Multisim®

Third Edition

Richard H. Berube
Community College of Rhode Island

Upper Saddle River, New Jersey
Columbus, Ohio

Editor in Chief: Stephen Helba
Editor: Dennis Williams
Development Editor: Kate Linsner
Production Editor: Rex Davidson
Design Coordinator: Diane Ernsberger
Cover Designer: Jeff Vanik
Cover art: Digital Vision
Production Manager: Pat Tonneman
Marketing Manager: Ben Leonard

This book was printed and bound by Courier Kendallville, Inc. The cover was printed by Phoenix Color Corp.

Pearson Education Ltd.
Pearson Education Singapore Pte. Ltd.
Pearson Education Canada, Ltd.
Pearson Education—Japan
Pearson Education Australia Pty. Limited
Pearson Education North Asia Ltd.
Pearson Educación de Mexico, S.A. de C.V.
Pearson Education Malaysia Pte. Ltd.

10 9 8 7 6 5 4 3 2 1

ISBN: 0-13-048788-0

Preface

Computer Simulated Experiments for Electric Circuits using Electronics Workbench Multisim®, Third Edition, is a unique and innovative laboratory manual that uses Multisim to simulate actual laboratory experiments on a computer. Computer simulated experiments do not require extensive laboratory facilities, and a computer provides a safe and cost-effective laboratory environment. Circuits can be modified easily with on-screen editing, and analysis results provide faster and better feedback than does a series of lab experiments using hardwired circuits.

The experiments are designed to help reinforce the theory learned in a dc and ac electric circuits course. By answering questions about the results of each experiment, students will develop a clearer understanding of the theory. Also, the interactive nature of these experiments encourages student participation in the learning process, which leads to more effective learning and a longer retention of the theoretical concepts. Experiments that involve a wide range of difficulty have been included so that experiments appropriate for a particular level of instruction can be selected. Experiments 8, 9, 17, 22, and 24–26 are more appropriate for more advanced students than most of the other experiments in this manual.

A series of troubleshooting problems is included at the end of many of the experiments to help students develop troubleshooting skills. In each troubleshooting problem, the parts bin has been removed to force the student to find a fault or a component value by making a series of circuit measurements using only the instruments provided. A solutions manual showing measured data, calculations, answers to the questions, and answers to the troubleshooting problems is available to instructors.

In the third edition, the circuits have been modified and some of the experiments have been changed for the Multisim 2001 circuit simulator. Also in the third edition, the "Preparation" section of each experiment has been renamed "Theory". This theory section has all of the technical information needed to do the calculations and answer the questions in the procedure section without referring to another textbook, and **makes it possible to use this manual as a combination text and lab manual, if desired**. Experiment 24 on series ac impedance from the previous edition has been combined with Experiment 21 in the new edition, and Experiment 25 on parallel ac admittance from the prior edition has been combined with Experiment 22 to eliminate the redundancy between these experiments. These modifications are an improvement over the second edition and will make these experiments work better with the new Multisim software.

Richard H. Berube
Email rberube@ccri.cc.ri.us

Contents

Introduction

Electronics Workbench Multisim® is similar to a workbench in a real laboratory environment, except that **circuits are simulated on a computer** and results are obtained more quickly. All of the components and instruments necessary to create and simulate mixed-mode analog and digital circuits on the computer screen are provided. Using a mouse, you can build a circuit in the central workspace, attach simulated test instruments, simulate actual circuit performance, and display the results on the test instruments. Because **circuit faults can be introduced without destroying or damaging actual components**, more extensive troubleshooting experiments can be performed. Also, faulty components that are deliberately introduced in a circuit simulated on a computer can help make it easier to find the faulty component in an actual circuit. The **Multisim Help menu** has all the information needed to get started using Multisim. Additional notes on using Multisim have been included in the Appendix at the end of this manual.

Each experiment includes a list of **Objectives**, a **Materials** list, a **Theory** section, **circuit diagrams**, a **Procedure** section, and a **Troubleshooting** section. The Procedure section requires you to record measured data, calculate expected values, and answer a series of questions designed to reinforce the theory. The Theory section provides all of the theory and equations needed to complete the procedure. The Troubleshooting section has a series of problems that require using the theory learned in the experiment to find a defective component or determine the value of a component.

The materials list and circuit diagrams make it possible to perform the experiments in a **hardwired laboratory** environment using actual circuit components. If you perform the experiments in a hardwired laboratory, wire the circuit from the circuit diagram and connect the instruments specified. If exact circuit component values are not available, use component values as close as possible to those listed on the parts list. After the circuit is wired and checked, turn on the power, record the data in the space provided, and answer the questions. Due to component tolerances, real laboratory data will not be exactly the same as data obtained on the computer simulation. If a closer correlation between the computer simulation results and the hardwired laboratory results is desired, change the component values in your computer simulation circuit to match the measured component values in your hardwired circuit. The circuit diagrams also make it possible to simulate a hardwired laboratory environment by wiring the circuits on the computer screen using the components from the parts bin.

The **CD-ROM** provided with this manual contains the circuits needed to perform the experiments in this manual, the troubleshooting circuits, and a student version of Electronics Workbench Multisim. This CD-ROM is write-protected; therefore, you will not be able to save circuit changes on the disk. (Use **save-as** to save circuit changes to another disk). If you wish to install the full educational version of Electronics Workbench Multisim onto your computer system, it can be obtained from **Electronics Workbench, 111 Peter Street, Suite 801, Toronto, Ontario, Canada M5V2H1**
(Tel 1-800-263-5552, or Fax 416-977-1818).

Part

I

Direct Current (DC) Circuits

The experiments in Part I involve the analysis of dc circuits. You will study the difference between voltage and current in dc circuits, Ohm's law and the concept of resistance, electrical power and the difference between power and energy, Kirchhoff's voltage law and resistors in series, Kirchhoff's current law and resistors in parallel, series-parallel resistance networks, the voltage and current divider rules, the nodal voltage and mesh current methods of circuit analysis, and Thevenin and Norton equivalent circuits.

The circuits for the experiments in Part I can be found on the enclosed disk in the PART1 subdirectory.

EXPERIMENT

1

Name________________________________

Date________________________________

Voltage and Current in DC Circuits

Objectives:

1. Investigate the meaning of electric **current**.
2. Learn how to use an **ammeter** to measure the current in an electric circuit.
3. Investigate the meaning of **voltage**.
4. Learn how to use a **voltmeter** to measure voltage in an electric circuit.
5. Demonstrate the two conditions necessary to allow a flow of electric charge (current) in an electric circuit.
6. Demonstrate the relationship between the voltage applied to an electric circuit and the current flow in an electric circuit.

Materials:

One 0–20 V variable dc voltage supply
One 0–20 V dc voltmeter
One 0–100 mA dc milliammeter
One 10 V, 1 W lamp

Theory:

Electric charge (Q) can be positive or negative. Electrons have negative charge and protons have positive charge. Unlike charges attract each other and like charges repel each other. Electric charge is measured in **coulombs**. A single electron has a charge of 1.6×10^{-19} coulombs. Therefore, it takes 6.28×10^{18} electrons to produce one coulomb of charge.

An **electric current (I)** is a measure of the rate of flow of electric charge. Mathematically, it can be represented as

$$I = \frac{Q}{t}$$

where Q is the electric charge in coulombs, t is time in seconds, and I is the electric current in **amperes**. An electric current of one ampere (1 A) is the rate of flow of an electric charge of one coulomb passing a point in an electric circuit in one second.

Voltage (V) is the driving force that causes the flow of charge (current) in an electric circuit. The voltage (electrical potential) between two points in an electric circuit is defined as the energy (W) produced per coulomb (C) of charge while moving the charge between the two points. Mathematically, it can be represented as

$$V = \frac{W}{Q}$$

where W is the energy in joules, Q is the charge in coulombs, and V is the voltage in **volts**. An electrical potential of one volt between two points in an electric circuit will produce one joule of electrical energy (work) while moving a charge of one coulomb between the two points. The electric current (I) in a circuit element is a function of the voltage (V) across the terminals of the circuit element. If the voltage is increased, the current increases, and if the voltage is decreased, the current decreases. However, an electric charge does not have to be in motion in order for an electrical potential (voltage) to exist. Electrical potential (voltage) is the potential to put charge in motion (do work), even if no charge is moved (no electrical current). If there is no current flow in a circuit element, the voltage (electrical potential) across that circuit element will be zero (0 V), unless the element is an open circuit.

An electric current is measured with an **ammeter**. To determine the value of an electric current, the circuit must be opened and the ammeter must be inserted in the path of the electric current so that the electric current will flow through the ammeter, as shown in Figure 1-1. **Never connect an ammeter across a circuit element.** This will cause an excessive amount of current to flow through the ammeter because of its low resistance.

Electrical potential (voltage) is measured with a **voltmeter**, and voltage is always measured between two points in an electric circuit. A voltmeter is always connected across an electric circuit element, as shown in Figure 1-1.

The two conditions necessary for an electric current to exist are a completely closed path and a source of electrical potential (voltage). A basic electric circuit consists of a voltage source, a load, and a closed current path, as shown in Figure 1-1. The battery (10 V) is the voltage source, the lamp (1 W/10 V) is the load, and a-b-c-d-a is the closed current path. If the closed current path is broken (open circuit) or the voltage of the source is reduced to zero, no current (I) will flow. The ground symbol in the circuit establishes point d as the circuit reference point (0 V). All of the ground points in a circuit are at the same electrical potential (voltage) and are therefore common points.

Note: The measured bulb current in a hardwired laboratory environment may not be exactly the same as the measured bulb current on the computer simulated circuit. Real light bulb voltage-current characteristics are nonlinear, while the bulb model in this circuit is a linear model.

Figure 1-1 DC Voltage and Current

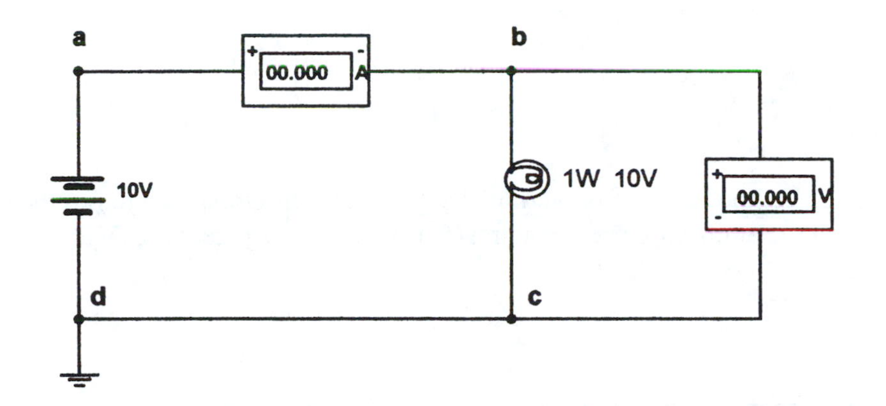

Procedure:

Step 1. Pull down the File menu and open FIG1-1. Click the On-Off switch to run the simulation. Record the voltage across the lamp terminals (V_{bc}) and the lamp current (I_{ab}).

V_{bc} = __________ I_{ab} = __________

Question: How did the voltage across the lamp (V_{bc}) compare with the dc source (battery) voltage? **Explain.**

Step 2. Based on the lamp current (I_{ab}), calculate the total charge (Q) passing through the lamp each second.

Step 3. Based on the voltage across the lamp (V_{bc}), calculate the energy required to pass 1 coulomb of charge through the lamp.

Step 4. Disconnect the battery from terminal a. Click the On-Off switch to run the simulation. Record the ammeter current reading (I_{ab}) and the voltage reading (V_{bc}).

I_{ab} = __________ V_{bc} = __________

Questions: What happened to the current reading when the circuit path was opened? **Explain.**

What happened to the voltage (V_{bc}) across the lamp when the circuit path was opened? **Explain.**

Step 5. Reconnect the battery to terminal a. Change the dc supply (battery) voltage to 0 V. Click the On-Off switch to run the simulation. Record the current (I_{ab}) reading.

I_{ab} = __________

Question: What happened to the current (I_{ab}) when the dc source (battery) voltage was reduced to zero? **Explain.**

Step 6. Change the dc supply (battery) voltage to 6 V. Run the simulation again. Record the new current reading.

I_{ab} = __________

Question: What happened to the current reading when the dc source (battery) voltage was changed to 6 V? How did the current reading in Step 6 compare with the current reading in Step 1? **Explain.**

Step 7. Disconnect the dc source (battery) from the circuit and rotate it until it is upside down. Reconnect the battery to the circuit at nodes a-d. Change the voltage to 10 V. Click the On-Off switch to run the simulation. Record the current (I_{ab}) and lamp voltage (V_{bc}).

V_{bc} = __________ I_{ab} = __________

Question: What happened to the current and voltage readings when the dc source (battery) was reversed? How did the current and voltage readings in Step 7 compare with the readings in Step 1? **Explain.**

Describe the difference between the way to connect a voltmeter to a circuit and the way to connect an ammeter to a circuit. What precaution must be taken when connecting an ammeter to a circuit? **Explain**.

EXPERIMENT

2

Name________________________________

Date________________________________

Ohm's Law–Resistance

Objectives:

1. Develop an understanding of the concept of resistance.
2. Learn how to use the multimeter to measure resistance.
3. Verify the validity of Ohm's law.
4. Use Ohm's law to determine the resistance of a resistor.
5. Verify the relationship between current and voltage for a constant resistance.
6. Verify the relationship between linear resistance and the slope of the V-I characteristic curve.

Materials:

One 0–20 V dc variable voltage supply
One multimeter
One 0–10 mA dc milliammeter
One 0–20 V dc voltmeter
Assorted resistors

Theory:

When there is an electrical current in any material, the moving charges collide with the atoms in the material, restricting the movement of the charges. The amount of restriction varies in different materials. This restrictive property, which produces an opposition to current flow, is called **resistance (R)** and is measured in **ohms (Ω)**. The resistance of a material is directly proportional to the length of the material (l) and inversely proportional to the cross sectional area of the material (A). Therefore, the resistance of a particular material can be represented mathematically as

$$R = \rho \frac{l}{A}$$

where ρ is the **resistivity** of the material in ohm-meters (Ω-m) and depends on the material, l is the length in meters (m), and A is the cross sectional area in square meters (m^2). A material with a very low resistivity (ρ) is called a **conductor**. A material with a very high resistivity (ρ) is called an **insulator**. Semiconductor materials have a resistivity that is between the resistivity of a conductor and an insulator.

For most materials, a change in temperature results in a change in resistivity. The change in resistivity per degree change in temperature is called the **temperature coefficient** (α) of the material. For some materials, an increase in temperature results in an increase in resistivity (**positive temperature coefficient**). For others, an increase in temperature results in a decrease in resistance (**negative temperature coefficient**). Conductors have a positive temperature coefficient, and semiconductors and insulators have a negative temperature coefficient.

Components that are specifically designed to have a certain resistance are called **resistors**. Resistors can be placed into two main categories: fixed and variable. A **fixed resistor** has a constant value that is set by the manufacturer and cannot be changed easily. **Variable resistors** are designed so that their resistance values can be varied easily. Variable resistors are often referred to as potentiometers and rheostats. The circuit symbol for a fixed resistor is shown in Figure 2-1 (R_1 and R_2).

Ohm's law gives the relationship between the voltage across the terminals of a linear resistance element and the current through it. Ohm's law states that the voltage across a linear resistance element is proportional to the current, and the proportionality constant is the resistance of the resistive element. Ohm's law can be expressed mathematically as follows:

$$V = R\,I$$

where R is the resistance in ohms (Ω), I is the current through the resistance element in amperes (A), and V is the voltage across the terminals of the resistance element in volts (V). An electric circuit element has a resistance of one ohm when a voltage of one volt across the terminals of the element causes a current of one ampere to flow. Ohm's law can also be expressed as $I = V/R$. This relationship shows that the current is inversely proportional to the resistance of the resistance element and proportional to the voltage across the resistance element. This means that a higher resistance produces more opposition to the flow of an electrical current and a higher voltage produces more current.

If you plot a graph of the current in a resistance element as a function of the voltage across it (V-I characteristic curve), you can determine whether the resistance of the element is **linear** or **nonlinear**. If the graph plots as a straight line, the resistance is linear; otherwise, it is nonlinear. If a resistance element is linear, the resistance is equal to the inverse of the slope of the line because

$$\text{slope} = \frac{\Delta I}{\Delta V}$$

and

$$R = \frac{V}{I} = \frac{\Delta V}{\Delta I} = \frac{1}{\text{slope}}$$

The circuit in Figure 2-1 shows how to measure the resistance of a resistor using a multimeter. The circuit in Figure 2-2 will be used to verify Ohm's law.

Figure 2–1 Measuring Resistance

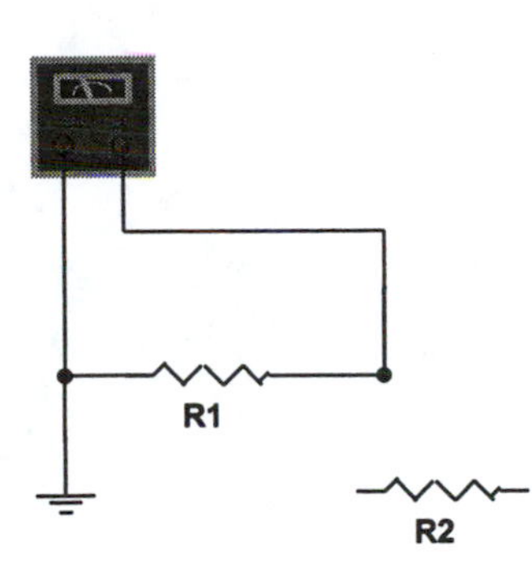

Figure 2-2 Ohm's Law—Resistance

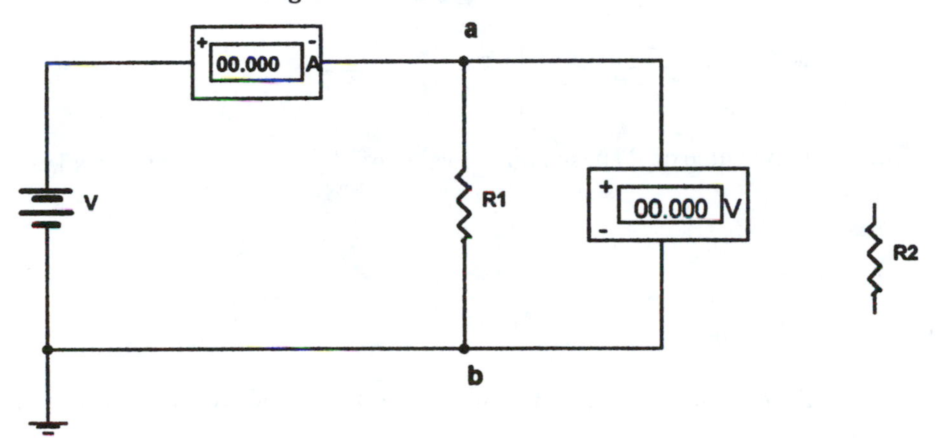

Procedure:

Step 1. Pull down the File menu and open FIG2-1. This circuit shows how to connect a multimeter to a resistor to measure the resistance. Make sure that Ω and dc(–) are selected on the multimeter. Click the On-Off switch to run the simulation. Record the resistance of R_1 measured by the multimeter. Next, remove R_1 and replace it with resistor R_2. Run the simulation again. Record the resistance of R_2 measured by the multimeter.

R_1 = ___________ R_2 = ___________

Step 2. Pull down the File menu and open FIG2-2. Click the On-Off switch to run the simulation. Record the voltage across resistor R_1 (V_{ab}) and the current through resistor R_1 (I_{ab}).

V_{ab} = ___________ I_{ab} = ___________

Step 3. Based on the measured values for V_{ab} and I_{ab}, and the value for R_1 measured in Step 1, verify Ohm's law for a linear resistance, $V = RI$.

Step 4. Replace R_1 with R_2. Run the simulation again. Record V_{ab} and I_{ab}.

V_{ab} = __________ I_{ab} = __________

Step 5. Calculate the value of R_2 based on the value of V_{ab} and I_{ab} using Ohm's law.

Questions: How did the calculated value of R_2 compare with the value of R_2 measured in Step 1?

For a constant voltage in Steps 2–4, what happened to the current (I_{ab}) when the resistance was increased? **Explain.**

Step 6. Change the value of the voltage source V to each value of V_{ab} in Table 2-1, run the simulations, and record each current value (I_{ab}) in resistor R_2.

Table 2-1

V_{ab} (V)	I_{ab} (mA)
0	
2	
4	
6	
8	
10	

Question: For a constant resistance in Step 6, what happened to the current (I_{ab}) when the voltage was increased? **Explain.**

Step 7. Replace resistor R_2 with resistor R_1. Repeat Step 6 and record the results in Table 2-2.

Table 2-2

V_{ab} (V)	I_{ab} (mA)
0	
2	
4	
6	
8	
10	

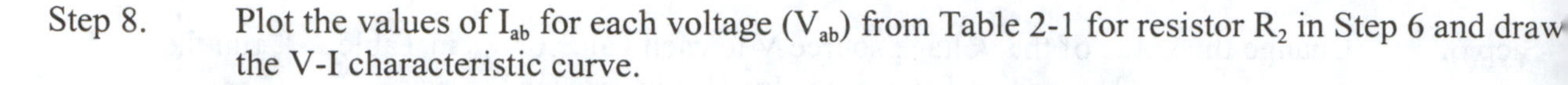

Step 8. Plot the values of I_{ab} for each voltage (V_{ab}) from Table 2-1 for resistor R_2 in Step 6 and draw the V-I characteristic curve.

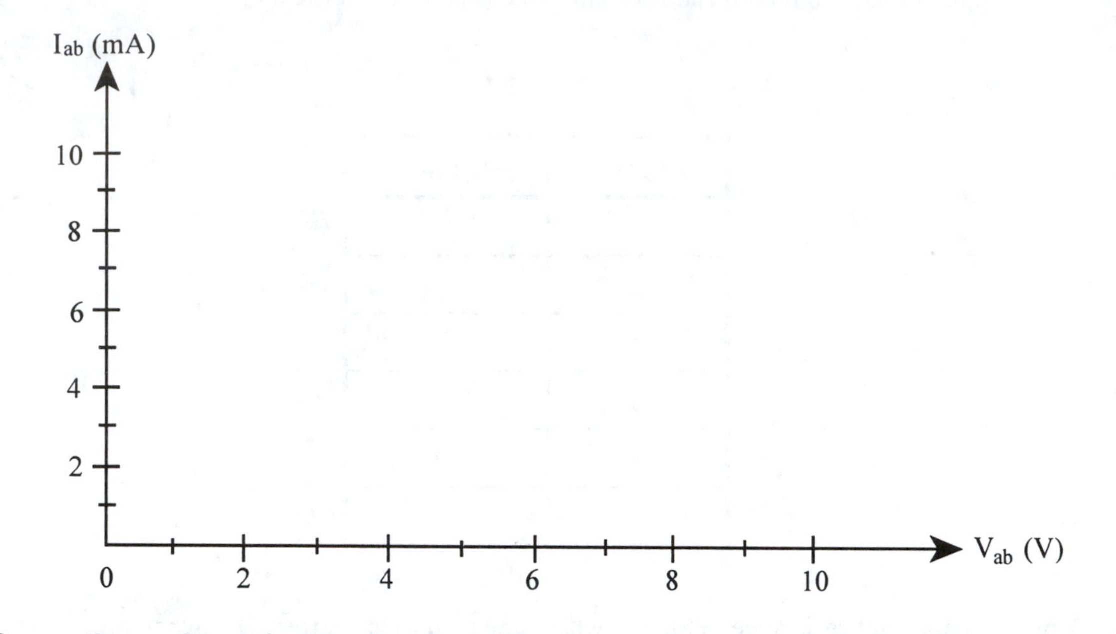

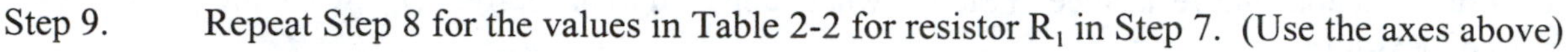

Step 9. Repeat Step 8 for the values in Table 2-2 for resistor R_1 in Step 7. (Use the axes above)

Question: Do the V-I characteristics plotted in Steps 8 and 9 indicate whether the resistors are linear or nonlinear? **Explain.**

Step 10. Calculate the value of each resistor based on the slope of the V-I characteristic curves plotted in Steps 8 and 9. Record the calculated resistance values on the curve plots.

Question: How did the resistance values calculated from the slope of the V-I graphs compare with the measured resistance values in Step 1? What is the relationship between the resistance and the slope?

Troubleshooting Problems

1. Pull down the File menu and open FIG2-3. Click the On-Off switch to run the simulation. Based on the voltage and current readings, determine the value of resistor R.

   R = __________

2. Pull down the File menu and open FIG2-4. Click the On-Off switch to run the simulation. Based on the current reading, determine the value of voltage V.

   V = __________

3. Pull down the File menu and open FIG 2-5. Click the On-Off switch to run the simulation. Based on the current reading, is resistor R_1 open or shorted?

4. Pull down the File menu and open FIG2-6. Click the On-Off switch to run the simulation. Based on the current reading, is resistor R_1 open or shorted?

EXPERIMENT

3

Name________________________________

Date________________________________

Electrical Power in DC Circuits

Objectives:

1. Determine the difference between power and energy.
2. Calculate the power dissipated in a light bulb based on the bulb current and voltage.
3. Calculate the resistance of a light bulb based on the power dissipated in the bulb and the bulb current.
4. Calculate the power dissipated in a light bulb based on the bulb voltage and resistance.
5. Demonstrate the relationship between power, voltage, and current.
6. Calculate the percent efficiency of power transfer based on load power output and source power input.
7. Demonstrate the relationship between load resistance and output power for a power source with internal resistance.
8. Determine the load resistance needed for maximum power output to the load.
9. Determine the load resistance needed for maximum efficiency.

Materials:

One 0–20 V dc voltage supply
One 0–20 V dc voltmeter
One 0–100 mA dc milliammeter
One 0–20 mA dc milliammeter
One 10 V, 1 W bulb
Resistors—100 Ω, 1 kΩ

Theory:

Electrical power is the rate of dissipation or generation of electrical energy (work) in joules per second. One **watt** of electrical power (P) is dissipated when one joule of work (W) is done in one second (1 watt = 1 joule/sec). Therefore,

$$P = \frac{W}{t}$$

One joule of energy (W) is required to raise one coulomb of electric charge (Q) through a potential difference of one volt (V). Therefore,

$$W = VQ$$

The electrical power (P) dissipated or generated in a circuit element in terms of the voltage (V) across the element and the current (I) flowing through the element can be found from

$$P = \frac{W}{t} = \frac{VQ}{t}$$

where current I = Q/t.

Therefore,

$$P = VI$$

When there is current flow in a resistance, energy is released in the form of heat. The electrical power (P) dissipated in a resistance R with current I flowing through it can be found from

$$P = VI = (RI)(I) = I^2R$$

The electrical power (P) dissipated in a resistance R with voltage (V) across it can be found from

$$P = VI = V\left(\frac{V}{R}\right) = \frac{V^2}{R}$$

The **percent efficiency** (η) of power transfer is the ratio of useful output power (P_o) to total input power (P_{in}) from the source. Therefore,

$$\eta = \frac{P_o}{P_{in}} \times 100\%$$

When transferring electrical energy from a source that has internal resistance (R_S) to a load (R_L), the maximum power output to the load occurs when

$$R_L = R_S$$

The maximum power output at the load does not coincide with the maximum efficiency (η). When a load resistance (R_L) is selected for maximum power output, equal power is dissipated in both the source resistance (R_S) and the load resistance (R_L). If high efficiency (η) and maximum voltage output are more important than maximum power output, the load resistance (R_L) should be much higher than the source resistance (R_S).

The circuit in Figure 3-1 will be used to demonstrate the measurement of the electrical power dissipated in a light bulb. The light bulb power rating is the power it will dissipate when the rated voltage is placed across the bulb terminals. The circuit in Figure 3-2 will be used to demonstrate the load resistance needed for maximum power transfer to the load. **Maximum power transfer** occurs when the load resistance (R_L) is equal to the output resistance of the source (R_S).

Figure 3-1 Electrical Power

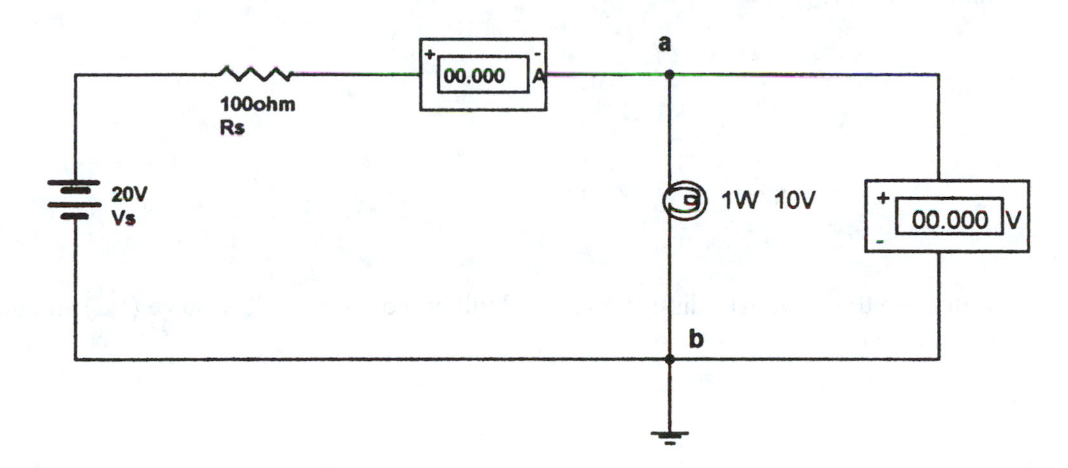

Figure 3-2 Maximum Power Transfer

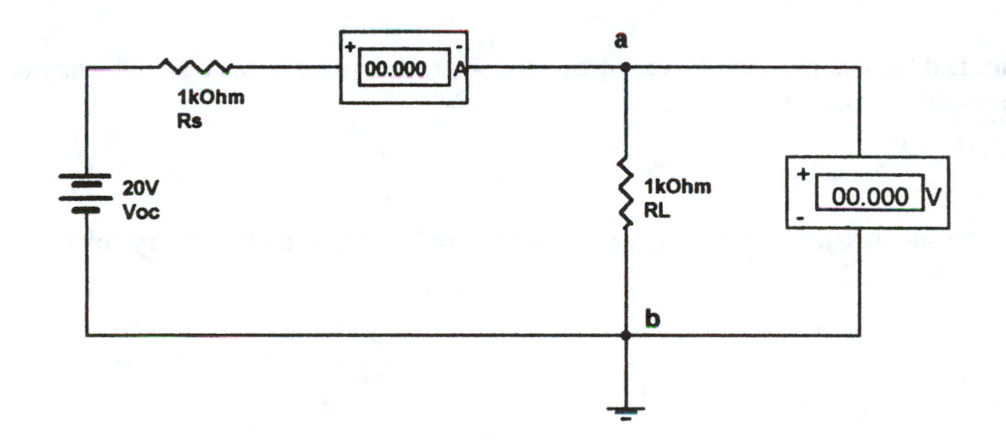

Procedure:

Step 1. Pull down the File menu and open FIG3-1. Click the On-Off switch to run the simulation. Record the bulb voltage (V_{ab}) and the bulb current (I_{ab}).

V_{ab} = ___________ I_{ab} = ___________

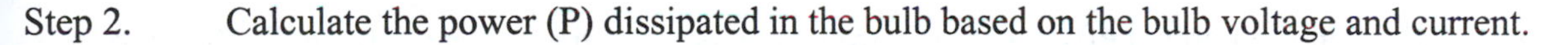

Step 2. Calculate the power (P) dissipated in the bulb based on the bulb voltage and current.

Question: Was the calculated power dissipated in the light bulb equal to the rated bulb power? **Explain.**

Step 3. Calculate the resistance (R) of the bulb based on the power (P) dissipated in the bulb and the bulb current (I_{ab}).

Step 4. Calculate the power (P) dissipated in the bulb based on the bulb voltage (V_{ab}) and bulb resistance (R).

Question: Did the two methods for calculating the power dissipated in the bulb in Steps 2 and 4 produce the same result?

Step 5. Calculate the power generated in the battery (P_{in}) based on the battery voltage (V_S) and current (I).

Step 6. Calculate the percent efficiency (η) of the power transfer from the battery to the bulb based on the power dissipated in the bulb (P_o) and the power generated in the battery (P_{in}).

Question: Based on the percent efficiency of power transfer from the battery to the bulb, did all of the power generated go to the bulb? If not, why not? Where did the rest of the power go?

Step 7. Pull down the File menu and open FIG3-2. Click the On-Off switch to run the simulation. Record the voltage (V_{ab}) and the current (I_{ab}) for each value of R_L in Table 3-1 below by changing the value of R_L and running the simulation each time.

Table 3-1

R_L (Ω)	V_{ab} (V)	I_{ab} (mA)	P_L (mW)	P_{in} (mW)	η (%)
100					
200					
500					
1000					
2000					
5000					
10,000					

Question: What happened to the load voltage (V_{ab}) and the load current (I_{ab}) as the load resistance (R_L) was increased? **Explain.**

Step 8. Calculate the load power (P_L) dissipated in R_L for each value of R_L and record your answers in Table 3-1.

Question: What happened to the load power (P_L) as the load resistance (R_L) was increased? **Explain.**

Step 9. Calculate the power generated in the battery (P_{in}) from the value of V_{oc} and I for each value of R_L and record your answers in Table 3-1.

Step 10. Calculate the percent efficiency (η) of the power transfer from the battery to the load based on the power dissipated in the load resistor (P_L) and the power generated in the battery (P_{in}) for each value of R_L and record your answers in Table 3-1.

Step 11. Plot the load power (P_L) as a function of load resistance (R_L). Note the value of R_L needed for maximum output power to the load on the curve plot.

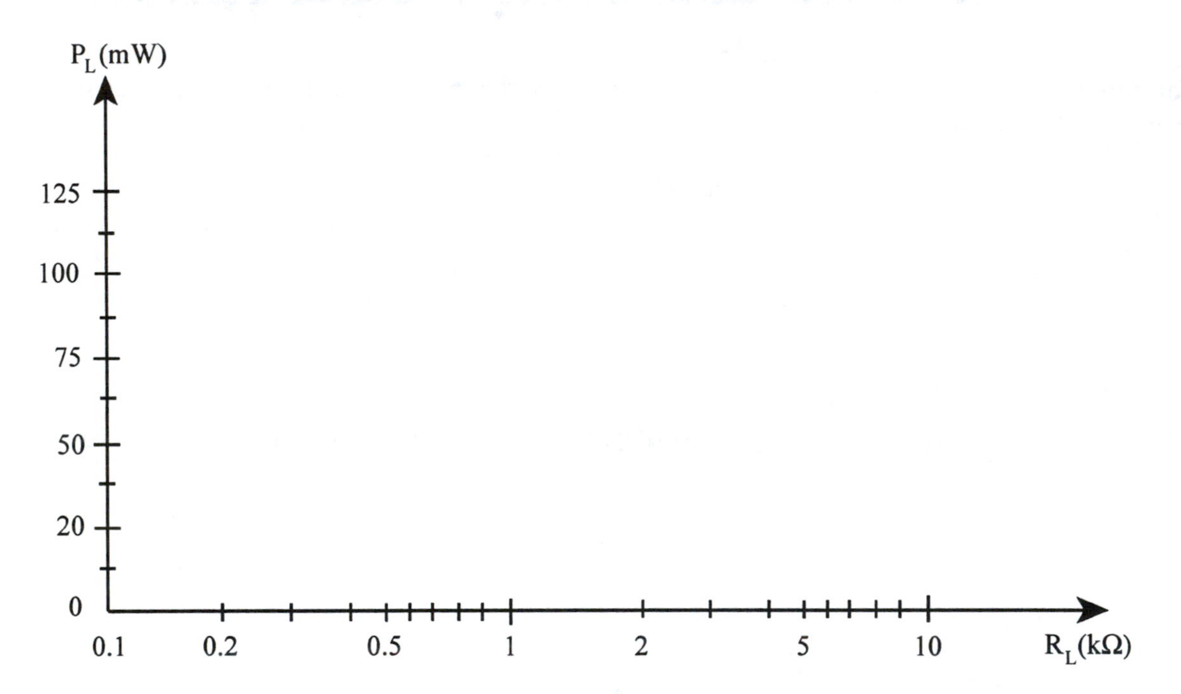

Question: What was the value of R_L needed for maximum power transfer from the source to the load? What was the relationship between this value of R_L and the source resistance (R_S)? **Explain.**

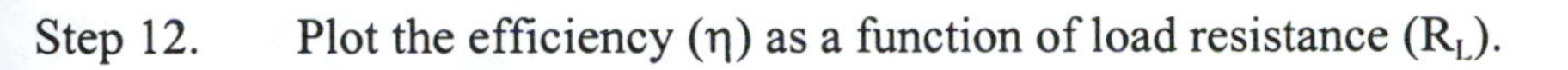

Step 12. Plot the efficiency (η) as a function of load resistance (R_L).

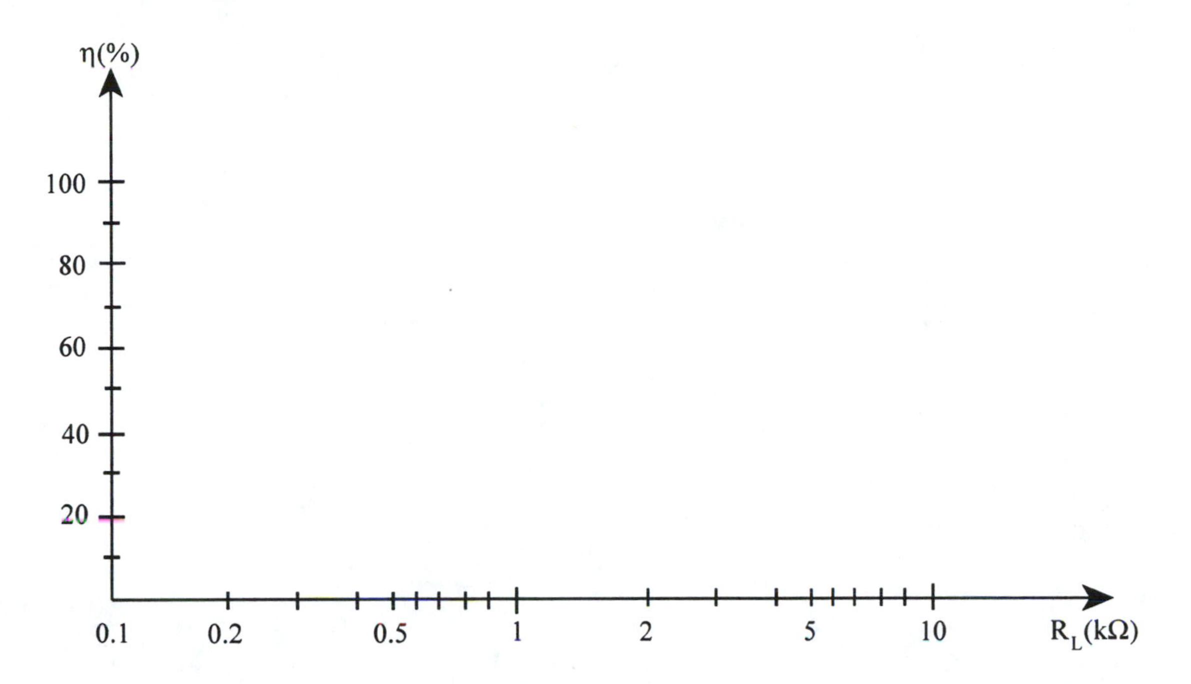

Question: What was the value of R_L needed for maximum efficiency? Was it the same value as needed for maximum output power? **Explain.**

EXPERIMENT

4

Name____________________

Date____________________

Resistors in Series—Kirchhoff's Voltage Law

Objectives:

1. Measure the equivalent resistance of a series resistance circuit and compare the measured value with the calculated value.
2. Determine the current in each resistor in a series resistance circuit.
3. Determine the voltage across each resistor in a series resistance circuit.
4. Determine the equivalent resistance of a series resistance circuit based on the circuit current and voltage.
5. Demonstrate Kirchhoff's voltage law.

Materials:

One 0–20 V dc voltage supply
One multimeter
Three 0–10 V dc voltmeters
One 0–5 mA dc milliammeter
Resistors—1 kΩ, 2 kΩ, and 3 kΩ

Theory:

Two or more circuit elements are considered to be in series if the electric current (I) is the same in every element. The equivalent resistance (R_{eq}) of a combination of resistors in series is equal to the sum of the individual resistors. Therefore, in Figure 4-1,

$$R_{eq} = R_1 + R_2 + R_3$$

Once the equivalent resistance of a series circuit is determined, you can solve for the common current (I) in all of the resistors from **Ohm's law** by dividing the voltage (V) applied across the series resistor combination by the equivalent resistance (R_{eq}) of the series resistors. Therefore,

$$I = \frac{V}{R_{eq}}$$

Kirchhoff's voltage law states that, around any closed path in an electric circuit, the algebraic sum of all the voltage drops must equal the algebraic sum of all the voltage rises. This means that the sum of the voltage drops across the series resistors in Figure 4-2 must equal the voltage applied across the series resistance combination. Therefore, using Kirchhoff's voltage law and Ohm's law,

$$V = V_1 + V_2 + V_3$$

where $V_1 = IR_1$, $V_2 = IR_2$, and $V_3 = IR_3$.

Figure 4-1 Resistors in Series—Equivalent Resistance

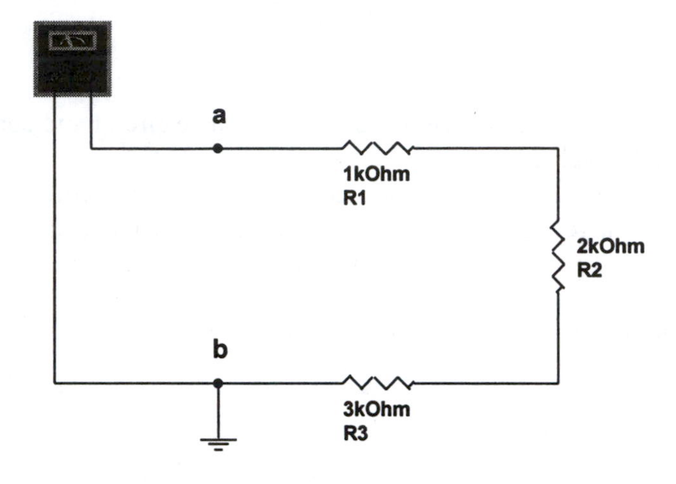

Figure 4-2 Resistors in Series—Kirchhoff's Voltage Law

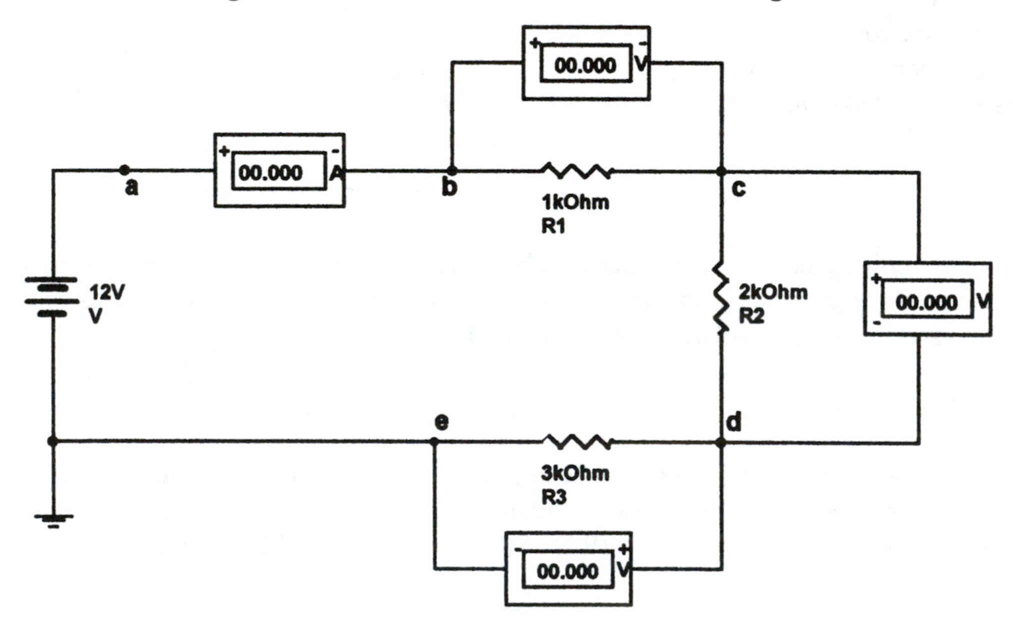

Procedure:

Step 1. Pull down the File menu and open FIG4-1. Using the multimeter to measure the equivalent resistance (R_{eq}) of the series circuit, click the On-Off switch to run the simulation. Record the measured resistance. Make sure that Ω is selected on the multimeter

R_{eq} = __________

Step 2. Calculate the equivalent resistance (R_{eq}) of the series resistors between terminals a-b.

Question: How did the calculated value for R_{eq} compare with the value measured in Step 1?

Step 3. Pull down the File menu and open FIG4-2. Click the On-Off switch to run the simulation. Record current I_{ab}, and voltages V_{bc}, V_{cd}, and V_{de}.

I_{ab} = __________

V_{bc} = __________ V_{cd} = __________ V_{de} = __________

Step 4. Based on the equivalent resistance (R_{eq}) calculated in Step 2 and the value of voltage source V, calculate the source current (I_{ab}).

Question: How did the calculated source current (I_{ab}) compare with the measured current I_{ab} in Step 3?

Step 5. Based on the current in R_1, use Ohm's law to calculate the voltage across resistor R_1 (V_{bc}).

Step 6. Based on the current in R_2, use Ohm's law to calculate the voltage across resistor R_2 (V_{cd}).

Step 7. Based on the current in R_3, use Ohm's law to calculate the voltage across resistor R_3 (V_{de}).

Question: How did the calculated values for voltages V_{bc}, V_{cd}, and V_{de} compare with the measured values in Step 3?

Step 8. Calculate the sum of voltages V_{bc}, V_{cd}, and V_{de}.

Questions: What was the relationship between voltage V and the sum of the voltages across the resistors (V_{bc}, V_{cd}, and V_{de}) in Step 8? Does this result confirm Kirchhoff's voltage law? **Explain.**

What is the relationship between the ratio of the voltages across resistors R_1, R_2, and R_3 and the ratio of the resistance values? **Explain.**

Troubleshooting Problems

1. Pull down the File menu and open FIG4-3. Click the On-Off switch to run the simulation. Based on the current and voltage readings, determine the resistance of R_1, R_2, and R_3.

   R_1 = __________ R_2 = __________ R_3 = __________

2. Pull down the File menu and open FIG 4-4. Click the On-Off Switch to run the simulation. Based on the current and voltage readings, determine the value of R_1.

   R_1 = __________

3. Pull down the File menu and open FIG4-5. Click the On-Off switch to run the simulation. Based on the current and voltage readings, determine the defective component and the defect.

   Defective component __________ Defect __________

4. Pull down the File menu and open FIG4-6. Click the On-Off switch to run the simulation. Based on the current and voltage readings, determine the defective component and the defect.

   Defective component __________ Defect __________

Name______________________________

Date______________________________

EXPERIMENT

5

Resistors in Parallel—Kirchhoff's Current Law

Objectives:

1. Measure the equivalent resistance of a parallel resistance circuit and compare the measured value with the calculated value.
2. Determine the current in each resistor in a parallel resistance circuit.
3. Determine the equivalent resistance of a parallel resistance circuit based on the circuit current and voltage.
4. Demonstrate Kirchhoff's current law.

Materials:

One 0–20 V dc voltage supply
One multimeter
One 0–20 V dc voltmeter
Four 0–5 mA dc milliammeters
Resistors—5 kΩ and 10 kΩ

Theory:

Two or more circuit components are considered to be in parallel if they are all connected between the same two nodes. A node is a point in a circuit where two or more components connect. The voltage is the same across parallel components. A parallel circuit provides more than one path for the current. Each current path in a parallel circuit is called a branch.

The inverse of the equivalent resistance of a combination of resistors in parallel is equal to the sum of the inverses of the individual resistors. Therefore, in Figure 5-1,

$$\frac{1}{R_{eq}} = \frac{1}{R_1} + \frac{1}{R_2} + \frac{1}{R_3}$$

You can solve for the equivalent resistance (R_{eq}) of the parallel circuit in Figure 5-2, using **Ohm's law** by dividing the voltage (V) applied across the parallel resistors by the total current entering the parallel resistance combination (I_{ab}). Therefore,

$$R_{eq} = \frac{V}{I_{ab}}$$

Kirchhoff's current law states that at any junction (node) in an electric circuit, the algebraic sum of all the currents entering the node must equal the algebraic sum of all the currents leaving the node. This means that the sum of the currents in the parallel resistors in Figure 5-2 must equal the total current entering the parallel resistance combination (I_{ab}). Therefore, using Kirchhoff's current law and Ohm's law,

$$I_{ab} = I_{bc} + I_{bd} + I_{be}$$

where $I_{bc} = V/R_1$, $I_{bd} = V/R_2$, and $I_{be} = V/R_3$.

Figure 5-1 Parallel Resistors—Equivalent Resistance

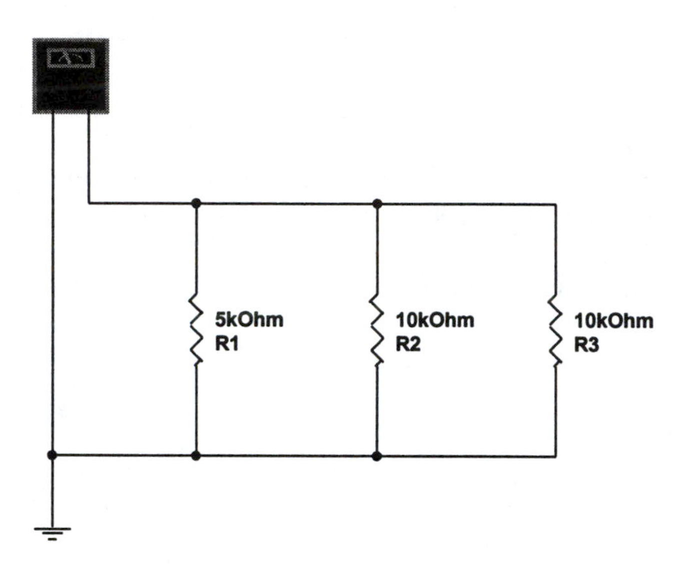

Figure 5-2 Parallel Resistors—Kirchhoff's Current Law

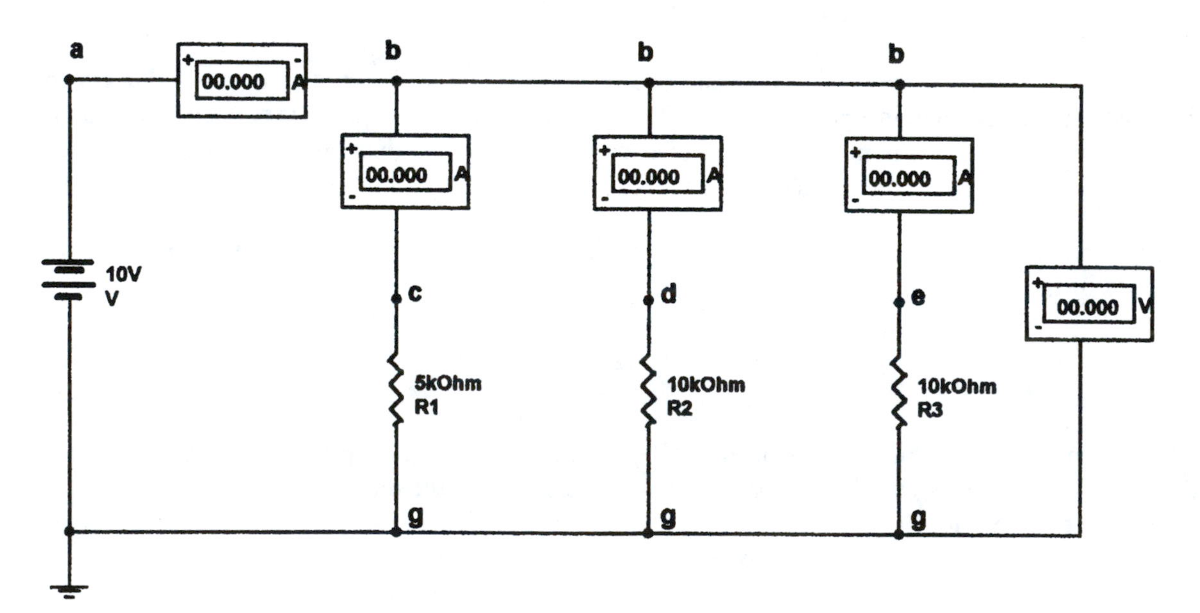

Procedure:

Step 1. Pull down the File menu and open FIG5-1. Using the multimeter to measure the equivalent resistance of the parallel circuit, click the On-Off switch to run the simulation. Record the equivalent resistance (R_{eq}). Make sure that Ω is selected on the multimeter.

R_{eq} = ________

Step 2. Calculate the equivalent resistance (R_{eq}) of the parallel resistors.

Question: How does the calculated value of the equivalent resistance (R_{eq}) compare with the value measured in Step 1?

Step 3. Pull down the File menu and open FIG5-2. Click the On-Off switch to run the simulation. Record currents I_{ab}, I_{bc}, I_{bd}, and I_{be}, and voltage V_{bg}.

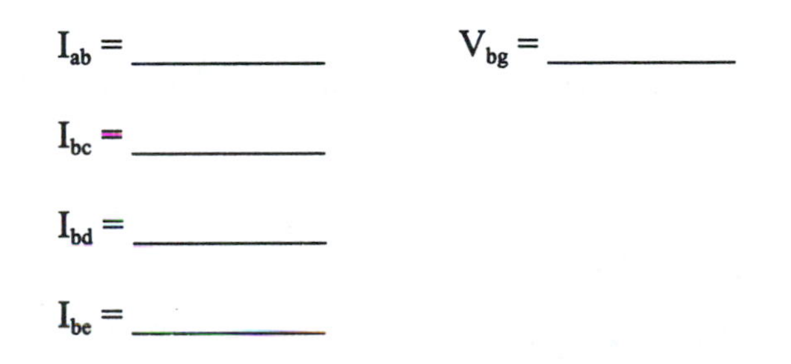

I_{ab} = ________ V_{bg} = ________

I_{bc} = ________

I_{bd} = ________

I_{be} = ________

Step 4. Based on the equivalent resistance (R_{eq}) calculated in Step 2 and the value of voltage source V, calculate the source current (I_{ab}).

Question: How did the calculated source current (I_{ab}) compare with the value measured in Step 3?

Step 5. Based on the voltage across R_1 and the resistance of R_1, calculate the current in resistor R_1 (I_{bc}).

Question: How did the calculated current in resister R_1 (I_{bc}) compare with the value measured in Step 3?

Step 6. Based on the voltage across R_2 and the resistance of R_2, calculate the current in resistor R_2 (I_{bd}).

Question: How did the calculated current in resister R_2 (I_{bd}) compare with the value measured in Step 3?

Step 7. Based on the voltage across R_3 and the resistance of R_3, calculate the current in resistor R_3 (I_{be}).

Question: How did the calculated current in resister R_3 (I_{be}) compare with the value measured in Step 3?

Step 8. Based on the circuit current (I_{ab}) and voltage V, calculate the equivalent resistance (R_{eq}) of the parallel circuit.

Question: How did the calculated equivalent resistance (R_{eq}) in Step 8 compare with the measured equivalent resistance in Step 1?

Step 9. Calculate the sum of currents I_{bc}, I_{bd}, and I_{be}.

Question: What was the relationship between current I_{ab} and the sum of the currents in the resistors (I_{bc}, I_{bd}, and I_{be}) in Step 9? Does this result confirm Kirchhoff's current law? **Explain.**

Troubleshooting Problems

1. Pull down the File menu and open FIG5-3. Click the On-Off switch to run the simulation. Based on the current and voltage readings, determine the resistance of R_1, R_2, and R_3.

   R_1 = __________ R_2 = __________ R_3 = __________

2. Pull down the File menu and open FIG5-4. Click the On-Off switch to run the simulation. Based on the current and voltage readings, determine the resistance of R_1.

   R_1 = __________

3. Pull down the File menu and open FIG5-5. Click the On-Off switch to run the simulation. Based on the current and voltage readings, determine the defective component and the defect.

   Defective component __________ Defect __________

4. Pull down the File menu and open FIG5-6. Click the On-Off switch to run the simulation. Based on the current and voltage readings, determine the defective component and the defect.

   Defective component __________ Defect __________

Name________________________________

EXPERIMENT

Date________________________________

Series-Parallel Circuits

Objectives:

1. Measure the equivalent resistance of a series-parallel circuit and compare your measured value with the calculated value.
2. Measure the current in each resistor in a series-parallel circuit and compare your measured values with the calculated values.
3. Measure the voltage across each resistor in a series-parallel circuit and compare your measured values with the calculated values.
4. Determine the equivalent resistance of a series-parallel circuit based on the circuit current and voltage.

Materials:

One 0–20 V dc voltage supply
One multimeter
Three 0–10 V dc voltmeters
Three 0–5 mA dc milliammeters
Resistors—2 kΩ and 4 kΩ

Theory:

A series-parallel circuit can be defined as one in which some portions of the circuit consist of circuit elements in series and other portions of the circuit consist of circuit elements in parallel. When two or more circuit elements are in series, all of the characteristics of a series circuit apply. When two or more circuit elements are in parallel, all of the characteristics of a parallel circuit apply. Therefore, to find the equivalent resistance of a series-parallel circuit, you must combine all of the series circuit elements to obtain the equivalent of the series portion of the circuit and combine all of the parallel circuit elements to obtain the equivalent of the parallel portion of the circuit until there is only one equivalent resistance left. **To learn how to combine series and parallel resistances, you must complete Experiments 4 and 5 before attempting this experiment.**

The equivalent resistance of the circuit in Figure 6-1 can be found from

$$R_{34} = R_3 + R_4$$

$$\frac{1}{R_{234}} = \frac{1}{R_2} + \frac{1}{R_{34}}$$

$$R_{eq} = R_1 + R_{234}$$

The equivalent resistance of the circuit in Figure 6-2 can be found from

$$R_{56} = R_5 + R_6$$

$$\frac{1}{R_{456}} = \frac{1}{R_4} + \frac{1}{R_{56}}$$

$$R_{3456} = R_3 + R_{456}$$

$$\frac{1}{R_{23456}} = \frac{1}{R_2} + \frac{1}{R_{3456}}$$

$$R_{eq} = R_1 + R_{23456}$$

To calculate the currents and voltages in all of the circuit elements in a series-parallel circuit, you must understand Ohm's law and Kirchhoff's current and voltage laws. These laws were studied in Experiments 2, 4, and 5. The circuits in Figures 6-3 and 6-4 are identical to the circuits in Figures 6-1 and 6-2 respectively, except that ammeters and voltmeters are connected for making current and voltage measurements and a voltage source is connected across terminals a-b in place of the multimeter.

Figure 6-1 Series-Parallel Resistors—Equivalent Resistance

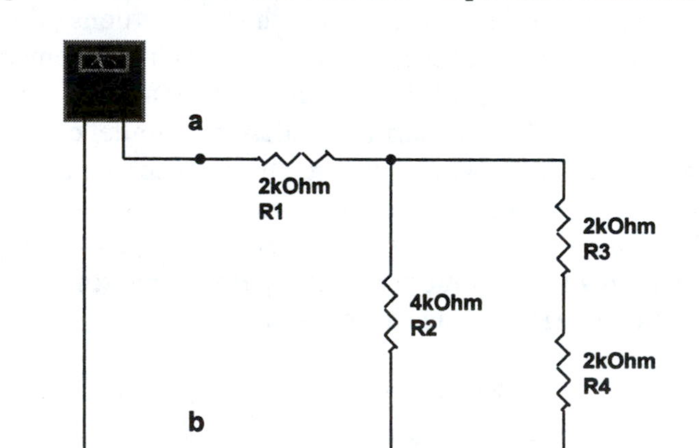

Figure 6-2 Series-Parallel Resistors—Equivalent Resistance

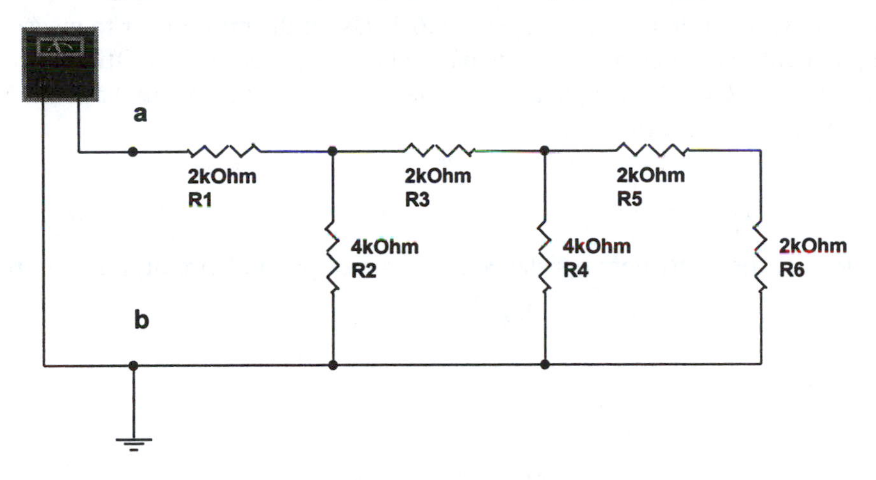

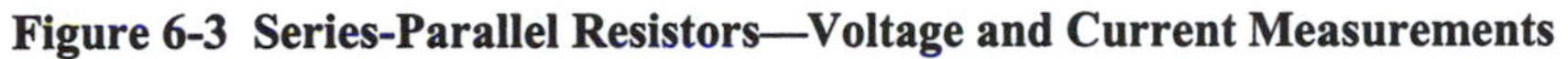

Figure 6-3 Series-Parallel Resistors—Voltage and Current Measurements

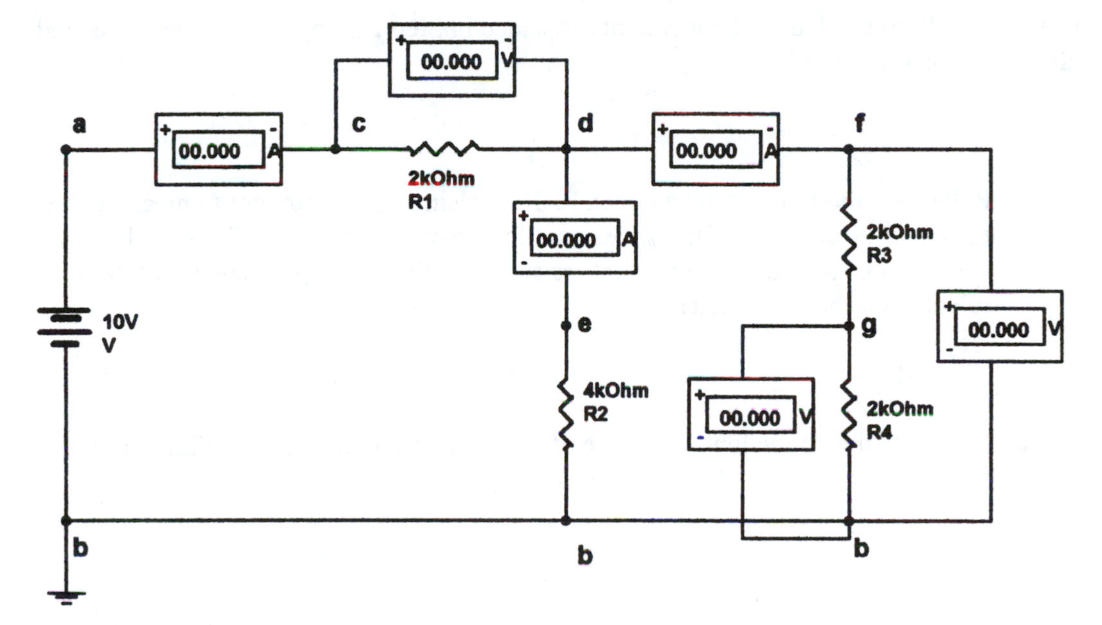

Figure 6-4 Series-Parallel Resistors—Voltage and Current Measurements

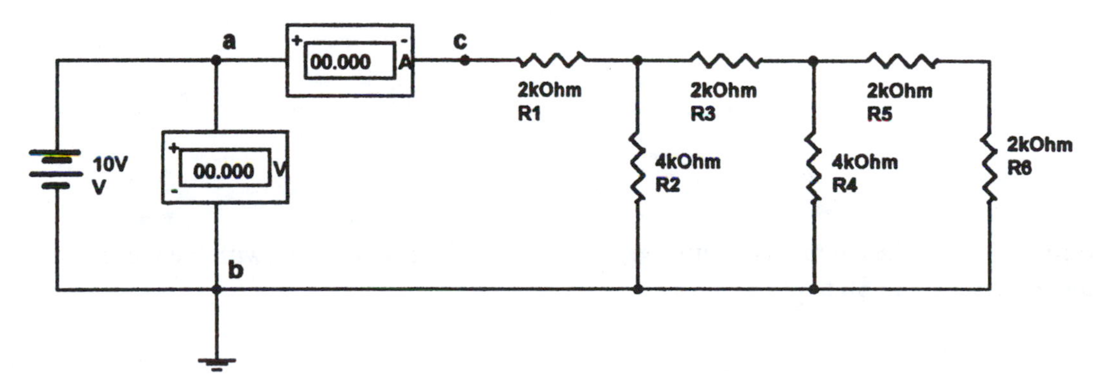

Procedure:

Step 1. Pull down the File menu and open FIG6-1. Using the multimeter to measure the equivalent resistance of the series-parallel circuit, click the On-Off switch to run the simulation. Record the equivalent resistance of the series-parallel circuit. Make sure Ω is selected on the multimeter.

R_{eq} = __________

Step 2. Calculate the equivalent resistance of the series-parallel circuit in Figure 6-1.

Question: How did the calculated equivalent resistance in Step 2 compare with the measured equivalent resistance in Step 1?

Step 3. Pull down the File menu and open FIG6-2. Using the multimeter to measure the equivalent resistance of the series-parallel circuit, click the On-Off switch to run the simulation. Record the equivalent resistance of the series-parallel circuit. Make sure Ω is selected on the multimeter.

R_{eq} = __________

Step 4. Calculate the equivalent resistance of the series-parallel circuit in Figure 6-2.

Question: How did the calculated equivalent resistance in Step 4 compare with the measured equivalent resistance in Step 3?

Step 5. Pull down the File menu and open FIG6-3. Click the On-Off switch to run the simulation. Record currents I_{ac}, I_{de}, and I_{df}, and voltages V_{cd}, V_{fb}, and V_{gb}.

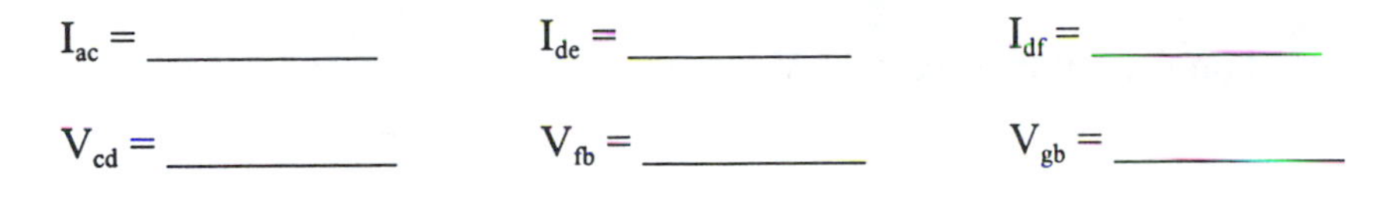

I_{ac} = ________ I_{de} = ________ I_{df} = ________

V_{cd} = ________ V_{fb} = ________ V_{gb} = ________

Question: What is the relationship between currents I_{ac}, I_{de}, and I_{df}? Does this relationship confirm Kirchhoff's current law?

Step 6. Based on the equivalent resistance (R_{eq}) calculated in Step 2 and the value of voltage source V, calculate the circuit current (I_{ac}).

Question: How did the calculated value for I_{ac} compare with the measured value in Step 5?

Step 7. Based on the value of current I_{ac} calculated in Step 6, calculate voltage V_{cd}.

Question: How did the calculated value for V_{cd} compare with the measured value in Step 5?

Step 8. Based on the value of voltage V_{cd} and source voltage V, use Kirchhoff's voltage law to calculate voltage V_{fb}.

Question: How did the calculated value for V_{fb} compare with the measured value in Step 5?

Step 9. Based on the value of voltage V_{fb}, calculate currents I_{de} and I_{df}.

Question: How did the calculated values for currents I_{de} and I_{df} compare with the measured values in Step 5?

Step 10. Based on current I_{df}, calculate voltage V_{gb}.

Question: How did the calculated value for voltage V_{gb} compare with the measured value in Step 5?

Step 11. Based on current I_{df}, calculate voltage V_{fg}.

Question: What is the relationship between voltages V_{fg}, V_{gb}, and V_{fb}?

Step 12. Based on the source voltage V and the measured circuit current I_{ac}, calculate the equivalent resistance (R_{eq}) of the series-parallel circuit in Figure 6-3.

Question: How did the calculated value of the equivalent resistance in Step 12 for the circuit in Figure 6-3 compare with the equivalent resistance measured in Step 1?

Step 13. Pull down the File menu and open FIG6-4. Click the On-Off switch to run the simulation. Record voltage V_{ab} and current I_{ac}.

V_{ab} = ________ I_{ac} = ________

Step 14. Based on the voltage and current measured in Step 13, calculate the equivalent resistance of the series-parallel circuit in Figure 6-4.

Question: How did the calculated value of the equivalent resistance in Step 14 for the circuit in Figure 6-4 compare with the equivalent resistance measured in Step 3?

Troubleshooting Problems

1. Pull down the File menu and open FIG6-5. Click the On-Off switch to run the simulation. Based on the current and voltage readings, determine the resistance of R_1, R_2, and R_3.

   R_1 = ________ R_2 = ________ R_3 = ________

2. Pull down the File menu and open FIG6-6. Click the On-Off switch to run the simulation. Based on the current reading, determine the voltage of voltage source V.

   V = ________

3. Pull down the File menu and open FIG6-7. Click the On-Off switch to run the simulation. Based on the current and voltage readings, determine the resistance of R_1, R_2, R_3, and R_4.

R_1 = __________ R_2 = __________

R_3 = __________ R_4 = __________

4. Pull down the File menu and open FIG6-8. Click the On-Off switch to run the simulation. Based on the current and voltage readings, determine the defective component and the defect.

Defective component __________ Defect __________

5. Pull down the File menu and open FIG6-9. Click the On-Off switch to run the simulation. Using the multimeter, determine the defective component and the defect.

Defective component __________ Defect __________

6. Pull down the File menu and open FIG6-10. Click the On-Off switch to run the simulation. Using the multimeter, determine the defective component and the defect.

Defective component __________ Defect __________

EXPERIMENT

7

Name________________________

Date________________________

Voltage and Current Divider Rules

Objectives:

1. Verify experimentally the voltage-divider rule.
2. Verify experimentally the current-divider rule.
3. Compare calculated and measured voltage and current values in a voltage-divider network.
4. Compare calculated and measured voltage and current values in a current-divider network.
5. Use the voltage-divider rule to calculate a voltage in a series resistive circuit.
6. Use the current-divider rule to calculate a current in a parallel resistive circuit.

Materials:

One 0–20 V dc voltage supply
Two 0–15 V dc voltmeters
Three 0–15 mA dc milliammeters
Resistors—2 kΩ and 3 kΩ

Theory:

The **voltage-divider rule** states that, in a series circuit, the ratio of voltage drops across resistances is the same as the ratio of the resistance values. Therefore, in Figure 7-1,

$$\frac{R_2}{R_{eq}} = \frac{V_{bg}}{V_{ag}} = \frac{V_{bg}}{V}$$

where $R_{eq} = R_1 + R_2$.

Also, in Figure 7-1,

$$\frac{R_2}{R_1} = \frac{V_{bg}}{V_{ab}}$$

The **current-divider rule** states that, in a parallel circuit, the ratio of any two branch currents is the same as the ratio of the two conductances of the branch resistances. Because conductance is the inverse of resistance ($G = 1/R$), the current divides inversely as the resistance. Therefore, in Figure 7-2,

$$\frac{I_{bd}}{I_{ab}} = \frac{G_2}{G_{eq}} = \frac{R_{eq}}{R_2}$$

where $G_{eq} = G_1 + G_2$

$$\text{and } \frac{1}{R_{eq}} = \frac{1}{R_1} + \frac{1}{R_2}$$

Also, in Figure 7-2,

$$\frac{I_{bd}}{I_{bc}} = \frac{G_2}{G_1} = \frac{R_1}{R_2}$$

Ohm's law is discussed in Experiment 2, and Kirchhoff's voltage and current laws are discussed in Experiments 4 and 5. Refer to the Theory sections of those experiments before performing Experiment 7.

Figure 7-1 Voltage-Divider Rule

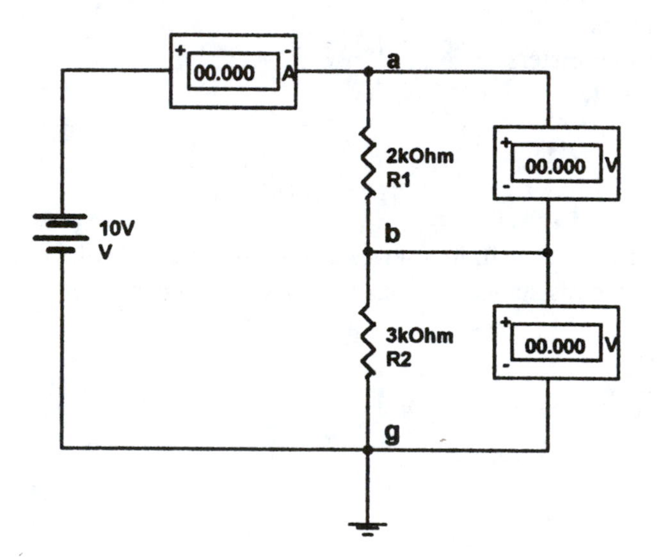

Figure 7-2 Current-Divider Rule

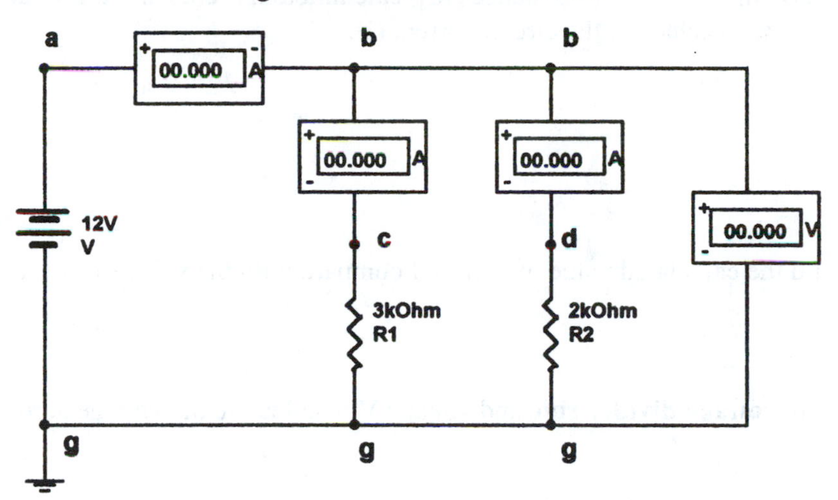

Procedure:

Step 1. Pull down the File menu and open FIG7-1. Click the On-Off switch to run the simulation. Record the circuit current (I) and voltages V_{ab} and V_{bg}.

I = __________

V_{ab} = __________ V_{bg} = __________

Question: What is the relationship between the voltage across R_2 (V_{bg}) and voltage across R_1 (V_{ab})? How does this compare with the relationship between the resistance of R_2 and the resistance of R_1? Does this verify the voltage-divider rule?

Step 2. Based on the resistance values in Figure 7-1, calculate the equivalent resistance (R_{eq}) of the series resistance circuit.

Question: What is the relationship between the voltage across R_2 (V_{bg}) and voltage V? How does this compare with the relationship between the resistance of R_2 and the equivalent resistance (R_{eq})? Does this verify the voltage-divider rule?

Step 3. Based on the equivalent resistance (R_{eq}) calculated in Step 2 and the value of V, use Ohm's law to calculate the circuit current (I).

Question: How did the calculated value of current I compare with the value of I measured in Step 1?

Step 4. **Use the voltage-divider rule** and voltage V to calculate the voltage across R_2 (V_{bg}).

Question: How did the calculated value of voltage V_{bg} compare with the value of V_{bg} measured in Step 1?

Step 5. **Use the voltage-divider rule** and voltage V to calculate the voltage across R_1 (V_{ab}).

Question: How did the calculated value of voltage V_{ab} compare with the value of V_{ab} measured in Step 1?

Step 6. Pull down the File menu and open FIG7-2. Click the On-Off switch to run the simulation. Record the value of currents I_{ab}, I_{bc}, and I_{bd}, and voltage V_{bg}.

I_{ab} = __________ I_{bc} = __________

I_{bd} = __________ V_{bg} = __________

Question: What is the relationship between branch current I_{bd} and branch current I_{bc}? How does this compare with the relationship between the resistance of R_2 and the resistance of R_1? Does this verify the current-divider rule?

Step 7. Based on the resistance values in Figure 7-2, calculate the equivalent resistance (R_{eq}) of the parallel resistance circuit.

Question: What is the relationship between branch current I_{bd} and circuit current I_{ab}? How does this compare with the relationship between the resistance of R_2 and the equivalent resistance (R_{eq}) of the parallel circuit? Does this verify the current-divider rule?

Step 8. **Use the current-divider rule** and circuit current I_{ab} to calculate the value of current I_{bc}.

Question: How did the calculated value of current I_{bc} compare with the value of I_{bc} measured in Step 6?

Step 9. **Use the current-divider rule** and circuit current I_{ab} to calculate the value of current I_{bd}.

Question: How did the calculated value of current I_{bd} compare with the value of I_{bd} measured in Step 6?

Troubleshooting Problems

1. Pull down the File menu and open FIG7-3. Click the On-Off switch to run the simulation. If the equivalent resistance of the series resistors is 6 kΩ, use the voltage-divider rule to determine the value of each resistor.

   R_1 = __________ R_2 = __________

2. Pull down the File menu and open FIG7-4. Click the On-Off switch to run the simulation. If the equivalent resistance of the parallel resistors is 2 kΩ, use the current-divider rule to determine the value of each resistor.

   R_1 = __________ R_2 = __________

EXPERIMENT

8

Name________________________

Date________________________

Nodal Voltage Circuit Analysis

Objectives:

1. Solve for the nodal voltage in a two-node circuit and compare your calculated value with the measured value.
2. Solve for the nodal voltages in a three-node circuit and compare your calculated values with the measured values.
3. Solve for the nodal voltages in a three-node circuit with a voltage source connected between two nodes and compare your calculated values with the measured values.

Materials:

Two 0–20 V variable voltage dc power supplies
Two 0–20 V dc voltmeters
Resistors–1 kΩ, 2 kΩ, and 6 kΩ

Theory:

The **nodal voltage method** of analyzing circuits requires using **Kirchhoff's current law** to determine the voltage at each circuit node (junction) with respect to a previously selected **reference node**, called the ground node. The steps required to solve for the nodal voltages are as follows:

1. Arbitrarily assign a reference node to be indicated with the ground symbol, normally at the bottom of the circuit.

2. Convert each voltage source (V_S) and series resistance (R_S) to its equivalent current source (I_S) in parallel with the resistance (R_S). The equivalent current source (I_S) is equal to V_S divided by R_S. This can be done only if there is a resistance in series with the source. (This step is optional, but it will make the calculations easier.)

3. Arbitrarily assign unknown voltage symbols (V_a, V_b, . . . , V_n) to each of the nodes, except for the reference node. The calculated nodal voltage values will be with respect to the reference node. If a voltage source is connected directly between two nodes without any resistors in the branch, assign a voltage symbol to one of the nodes and represent the other nodal voltage in terms of the first nodal voltage and the source voltage. If a voltage source is connected directly between a node and the reference node without any resistors in the branch, then the nodal voltage is equal to the source voltage.

4. Arbitrarily assign current directions to each circuit branch in which there is no current source. Normally the current directions are toward the reference node.

5. Apply Kirchhoff's current law to each node (except the reference node) and represent the branch currents in terms of the unknown nodal voltages and branch resistances. You will obtain one equation at each node (except the reference node). Therefore, if there are n nodes, you will obtain n – 1 equations. If there is a voltage source between two nodes without any resistance in the branch, you must treat the two nodes as one supernode.

6. Solve the resulting equations simultaneously to obtain the nodal voltage values.

Before completing the experiment, you may need to review the nodal method of circuit analysis in your textbook. It would also be helpful to review source transformation (converting equivalent voltage sources to equivalent current sources).

In this experiment you will analyze the two-node circuit in Figure 8-1, the three-node circuit in Figure 8-2, and the three node circuits with a voltage source connected between two nodes in Figures 8-3 and 8-4.

> ***Note*: This experiment is for advanced students and may be skipped without a loss in continuity.**

Figure 8-1 Nodal Analysis—Two-Node Circuit

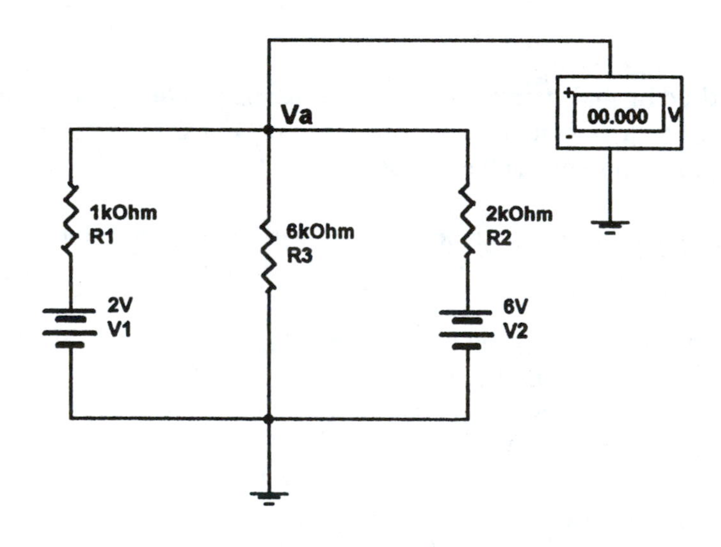

Figure 8-2 Nodal Analysis—Three-Node Circuit

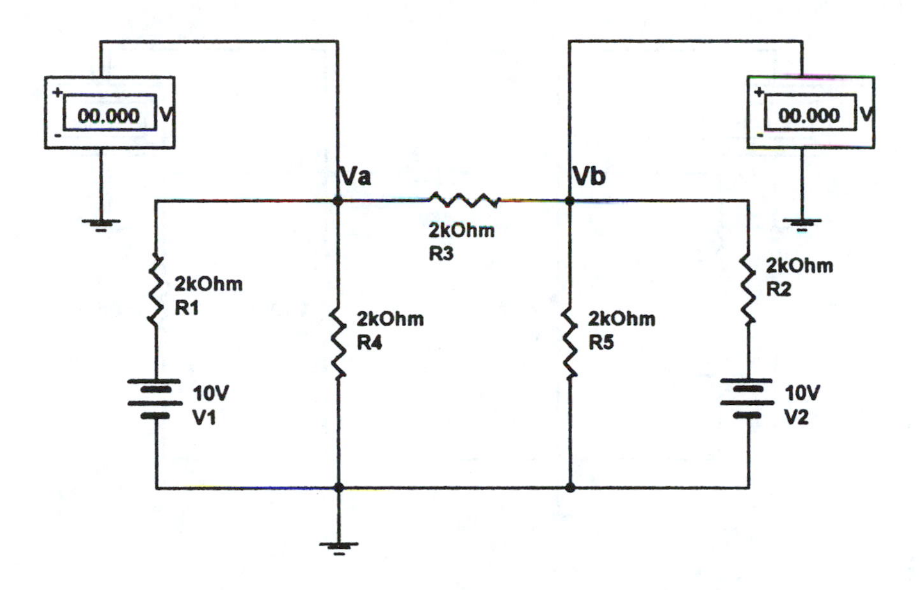

Figure 8-3 Nodal Analysis—Three-Node Circuit

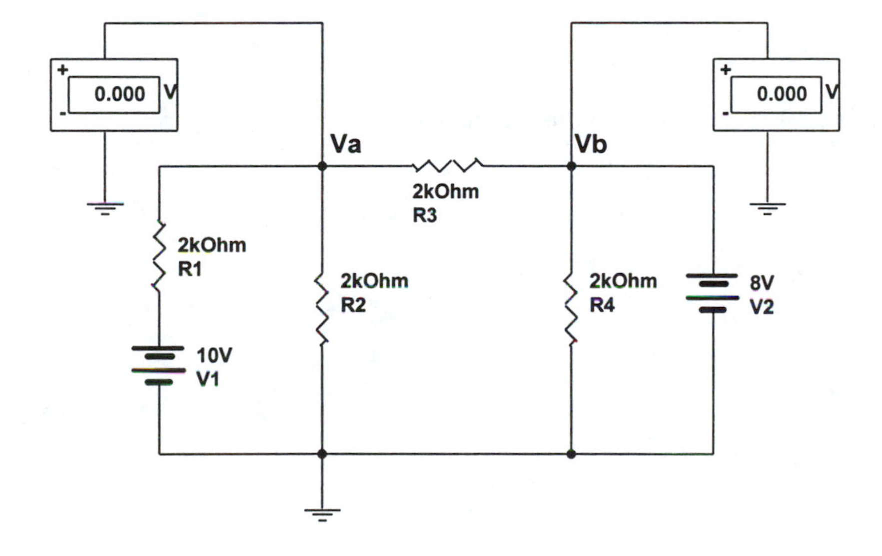

Figure 8-4 Nodal Analysis—Three-Node Circuit

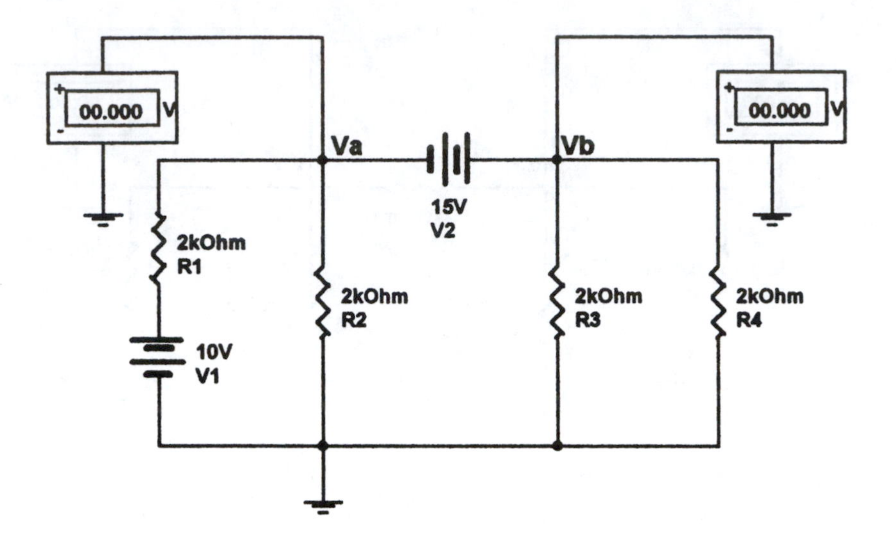

Procedure:

Step 1. Pull down the File menu and open FIG8-1. Click the On-Off switch to run the simulation. Record nodal voltage V_a.

V_a = ___________

Step 2. Solve for nodal voltage V_a in Figure 8-1.

Question: How did your calculated value for V_a compare with the measured value in Step 1?

Step 3. Pull down the File menu and open FIG8-2. Click the On-Off switch to run the simulation. Record nodal voltages V_a and V_b.

V_a = __________ V_b = __________

Step 4. Solve for nodal voltages V_a and V_b in Figure 8-2.

Question: How did your calculated values for V_a and V_b compare with the measured values in Step 3?

Step 5. Pull down the File menu and open FIG8-3. Click the On-Off switch to run the simulation. Record nodal voltages V_a and V_b.

V_a = __________ V_b = __________

Step 6. Solve for nodal voltages V_a and V_b in Figure 8-3.

Question: How did your calculated values for V_a and V_b compare with the measured values in Step 5?

Step 7. Pull down the File menu and open FIG8-4. Click the On-Off switch to run the simulation. Record nodal voltages V_a and V_b.

V_a = __________ V_b = __________

Step 8. Solve for nodal voltages V_a and V_b in Figure 8-4.

Question: How did your calculated values for V_a and V_b compare with the measured values in Step 7?

Name________________________________

Date________________________________

Mesh Current Circuit Analysis

Objectives:

1. Solve for the mesh currents in a two-loop circuit.
2. Solve for the mesh currents in a three-loop circuit.
3. Based on the mesh current values, determine the current in each branch and compare your calculated values with the measured values.

Materials:

Two 0–20 V variable voltage dc power supplies
Six 0–10 mA dc milliammeters
Resistors—1 kΩ

Theory:

The **mesh current method** of analyzing circuits requires using **Kirchhoff's voltage law** to sum voltages around a closed path in terms of the mesh (loop) current variables. The steps required to solve for the mesh currents are as follows:

1. Arbitrarily assign a clockwise current to each interior closed loop (mesh) and label it with an unknown variable ($I_1, I_2, \ldots, I_n$). If a current source is in a loop that is not in common with another loop, then the mesh current is equal to the source current, taking direction into account. If a current source is in a loop that is in common with another loop, the sum of the mesh currents is equal to the source current, taking direction into account.

2. Assign voltage polarities across all of the resistors in the circuit based on the mesh current directions. For a resistor that is in common with two loops, the polarity of the voltage due to each mesh current should be indicated on the appropriate side of the resistor.

3. Apply Kirchhoff's voltage law to each loop to develop the mesh current equations. You will obtain one equation for each loop, except for loops where there is a current source. If there are current sources in the circuit, you will obtain one equation from each source.

4. Solve the resulting equations simultaneously to obtain the mesh current values.

5. Solve the branch currents from the mesh currents. A branch that is in common with more than one loop has a branch current that is determined by algebraically combining the mesh currents that are common to the branch.

Before completing this experiment, you may need to review the mesh current method of circuit analysis in your textbook.

In this experiment you will analyze the two-loop circuit in Figure 9-1 and the three-loop circuit in Figure 9-2.

Figure 9-1 Mesh Current Analysis of a Two-Loop Circuit

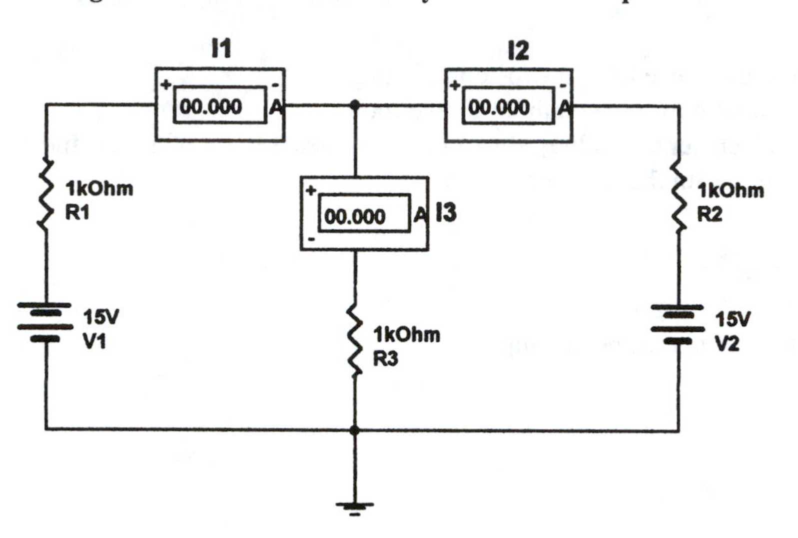

Figure 9-2 Mesh Current Analysis of a Three-Loop Circuit

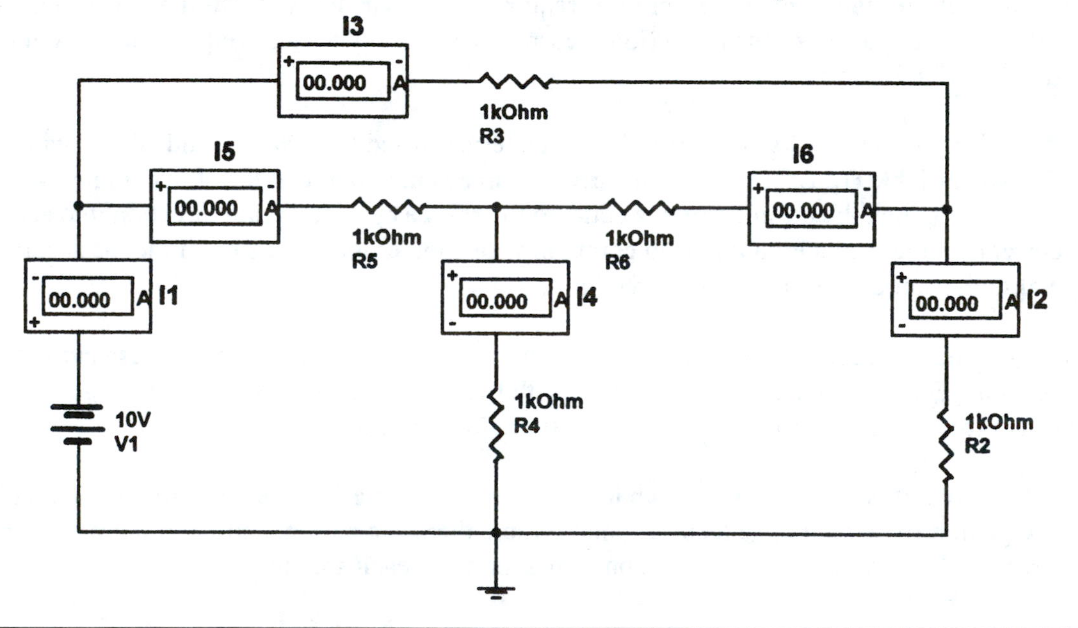

> ***Note*: This experiment is for advanced students and may be skipped without a loss in continuity.**

Procedure:

Step 1. Pull down the File menu and open FIG9-1. Click the On-Off switch to run the simulation. Record currents I_1, I_2, and I_3.

I_1 = __________ I_2 = __________ I_3 = __________

Step 2. Solve for the mesh currents in Figure 9-1. Draw the circuit and label the mesh currents.

Step 3. From the mesh current values, calculate the values of the branch currents I_1, I_2, and I_3.

Question: How did your calculated values for branch currents I_1, I_2, and I_3 compare with the measured values in Step 1?

Step 4. Pull down the File menu and open FIG9-2. Click the On-Off switch to run the simulation. Record currents I_1, I_2, I_3, I_4, I_5, and I_6.

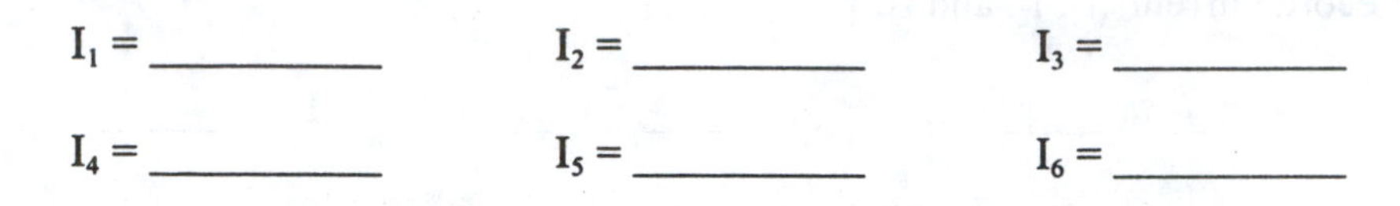

I_1 = __________ I_2 = __________ I_3 = __________

I_4 = __________ I_5 = __________ I_6 = __________

Step 5. Solve for the mesh currents in Figure 9-2. Draw the circuit and label the mesh currents.

Step 6. From the mesh current values, calculate the values of branch currents I_1, I_2, I_3, I_4, I_5, and I_6.

Question: How did your calculated values for the branch currents compare with the measured values in Step 4?

Name_______________________________

EXPERIMENT

10

Date_______________________________

Thevenin and Norton Equivalent Circuits

Objectives:

1. Determine the Thevenin equivalent circuit for a known network.
2. Determine the Norton equivalent circuit for a known network.
3. Determine the validity of Thevenin's theorem.
4. Determine the validity of Norton's theorem.
5. Determine the Thevenin equivalent circuit for an unknown network.
6. Determine the Norton equivalent circuit for an unknown network.

Materials:

One 0–20 V dc voltage supply
One 0–10 V dc voltmeter
One 0–1 mA dc milliammeter
Resistors–5 kΩ, 10 kΩ, and 15 kΩ

Theory:

Thevenin's Theorem

Thevenin's theorem is a very powerful tool for simplifying a linear two-terminal network of fixed resistances and voltage sources by replacing the network with a single voltage source in series with a single resistor. The single voltage source is called the **Thevenin voltage (V_{TH})** and is equal to the **open circuit voltage (V_{oc})** at the terminals (a-b) of the original network, as shown in Figure 10-1. The single resistor is called the **Thevenin resistance (R_{TH})** and is equal to the open circuit voltage (V_{oc}) at the terminals divided by the short circuit current (I_{sc}) between the terminals of the original network. Therefore,

$$V_{TH} = V_{oc}$$

and

$$R_{TH} = \frac{V_{oc}}{I_{sc}}$$

The **short circuit current (I_{sc})** is measured by connecting an ammeter between the terminals (a-b) of the original network and recording the current reading, as shown in Figure 10-2. (*Note:* An ammeter has a very low internal resistance and can be considered to be a short.) The short circuit current (I_{sc}) is calculated by drawing a short between the terminals (a-b) of the original network and calculating the current in the short.

An alternate method of determining the Thevenin resistance (R_{TH}) is by replacing all voltage sources with a short circuit and all current sources with an open in the original network and determining the equivalent resistance across the terminals. This equivalent resistance is equal to the Thevenin resistance (R_{TH}).

The best way to determine the Thevenin resistance (R_{TH}) for an unknown network in the laboratory is to connect a variable resistor across the terminals of the unknown network and vary the resistance until the terminal voltage is equal to one-half the open circuit voltage (V_{oc}). Then measure the value of the variable resistor, which will be equal to the Thevenin resistance (R_{TH}).

The strength of the Thevenin theorem lies in the fact that, although the Thevenin equivalent circuit is not the original circuit, it acts like the original circuit in terms of the voltage and current at the terminals.

Norton's Theorem

Norton's theorem states that any linear two-terminal network of fixed resistances and voltage sources may be replaced with a single current source in parallel with a single resistor. The single current source is called the **Norton current (I_N)** and is equal to the short circuit current (I_{sc}) between the terminals of the original network. The single resistor is called the **Norton resistance (R_N)** and is equal to the Thevenin resistance (R_{TH}) in the Thevenin equivalent circuit. It is found by following the same procedure that was used to find the Thevenin resistance (R_{TH}).

When a resistance (R_L) is connected between the terminals a-b of the original network, as shown in Figure 10-3, the voltage across the terminals (V_{ab}) will be the same as the voltage across the terminals of the Thevenin equivalent circuit if the same value resistance (R_L) is connected across the terminals of the Thevenin equivalent circuit. The same statement can be made for the Norton equivalent circuit.

Figure 10-1 Determining the Open Circuit Voltage

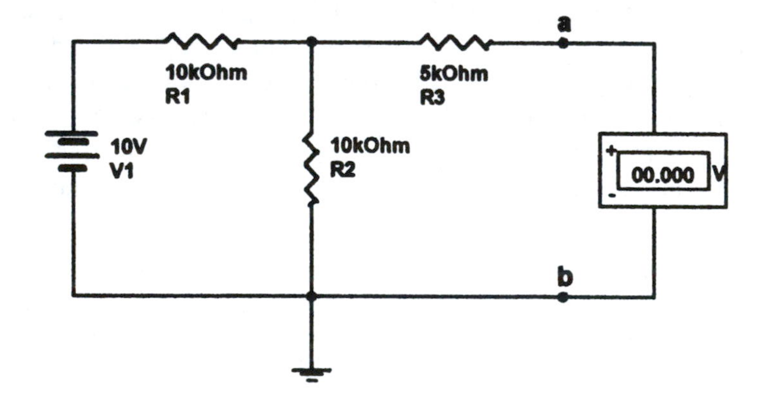

Figure 10-2 Determining the Short Circuit Current

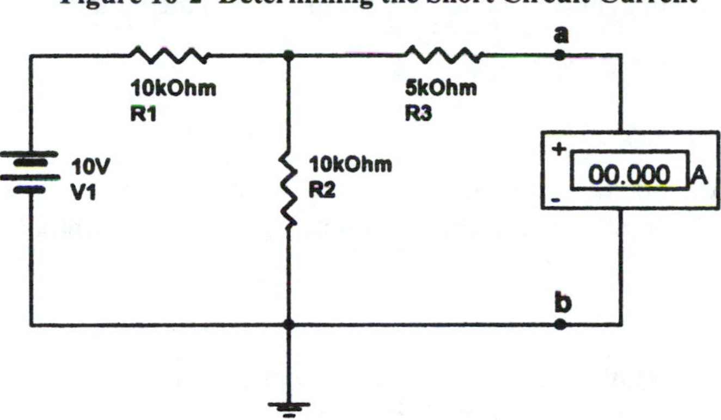

Figure 10-3 Determining Voltage V_{ab} with R_L Added

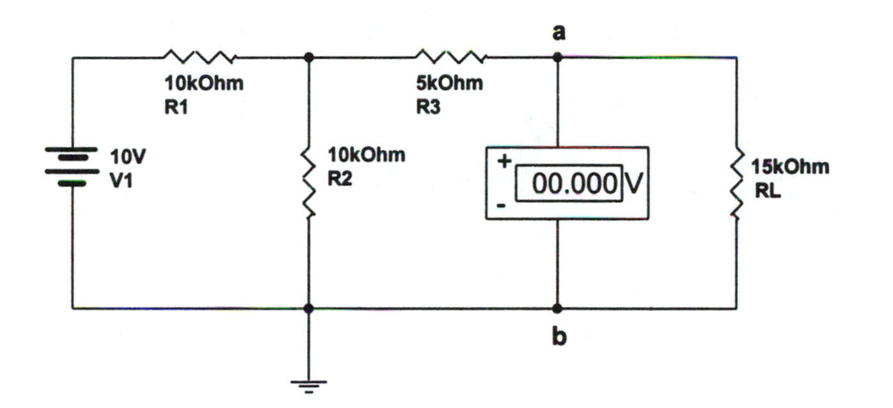

Procedure:

Step 1. Pull down the File menu and open FIG10-1. Click the On-Off switch to run the simulation. Record the open circuit voltage (V_{oc}) across terminals a-b.

V_{oc} = ___________

Step 2. Based on the circuit component values in Figure 10-1, calculate the expected value of the voltage across terminals a-b (V_{oc}). Draw the circuit and do calculations in the space below.

Question: How did your calculated value for V_{oc} compare with the value measured in Step 1?

Step 3. Pull down the File menu and open FIG10-2. Click the On-Off switch to run the simulation. Record the short circuit current (I_{sc}) between terminals a-b.

I_{sc} = __________

Step 4. Based on the circuit component values in Figure 10-2, calculate the expected value of the short circuit current (I_{sc}). Hint: Redraw the circuit and replace the ammeter with a short circuit.

Question: How did your calculated value for I_{sc} compare with the value measured in Step 3?

Step 5. Based on the measured values for V_{oc} and I_{sc}, calculate the Thevenin voltage (V_{TH}) and the Thevenin resistance (R_{TH}).

Step 6. Based on the values calculated in Step 5, draw the Thevenin equivalent circuit.

Step 7. Draw the circuit in Figure 10-2 with the milliammeter removed and voltage source V_1 replaced with a short circuit. Use this circuit to calculate the Thevenin resistance (R_{TH}) across terminals a-b.

Question: How did your calculated value for R_{TH} in Step 7 compare with the value calculated in Step 5?

Step 8. Based on the measured values for V_{oc} and I_{sc}, calculate the Norton current source (I_N) and the Norton resistance (R_N).

Step 9. Based on the values calculated in Step 8, draw the Norton equivalent circuit.

Step 10. Pull down the File menu and open FIG10-3. Click the On-Off switch to run the simulation. Record the value of voltage V_{ab}.

V_{ab} = __________

Step 11. Based on the circuit in Figure 10-3, calculate the expected value of voltage V_{ab}. Draw the circuit and do your calculations in the space below.

Question: How did your calculated value for V_{ab} compare with the value measured in Step 10?

Step 12. Draw the circuit in Figure 10-3 with the section to the left of terminals a-b replaced with the Thevenin equivalent circuit in Step 6. Use this circuit to solve for voltage V_{ab}.

Question: How did your answer for V_{ab} in Step 12 compare with your answer for V_{ab} in Steps 10 and 11? Was the Thevenin equivalent circuit equivalent to the original circuit? **Explain.**

Step 13. Draw the circuit in Figure 10-3 with the section to the left of terminals a-b replaced with the Norton equivalent circuit in Step 9. Use this circuit to solve for voltage V_{ab}.

Question: How did your answer for V_{ab} in Step 13 compare with your answer for V_{ab} in Steps 10 and 11? Was the Norton equivalent circuit equivalent to the original circuit? **Explain.**

Troubleshooting Problems

1. Pull down the File menu and open FIG10-4. Make the appropriate measurements using the voltmeter and the milliammeter to determine the Thevenin and Norton equivalent circuits for the unknown circuit inside the box labeled Circuit 1. Draw the Thevenin and Norton equivalent circuits with the correct component values in the space below.

2. Pull down the File menu and open FIG10-5. Make the appropriate measurements using the voltmeter and variable resistor to determine the Thevenin and Norton equivalent circuits for the unknown circuit inside the box labeled Circuit 2. Draw the Thevenin and Norton equivalent circuits with the correct component values in the space below.

Part

Instruments

The experiments in Part II involve an investigation of the loading effects of dc instruments and a study of the function generator and the oscilloscope. The oscilloscope is one of the most important and versatile electrical measuring instruments available.

The circuits for the experiments in Part II can be found on the enclosed disk in the PART2 subdirectory.

Experiment 11

Name________________________________

Date________________________________

Loading Effects of DC Instruments

Objectives:

1. Demonstrate how the internal resistance of a dc voltmeter can introduce voltage measurement errors.
2. Demonstrate how the internal resistance of a dc ammeter can introduce current measurement errors.
3. Show how to reduce voltage measurement errors.
4. Show how to reduce current measurement errors.

Materials:

One 0–20 V dc voltage supply
One 0–10 V dc voltmeter (low resistance)
One 0–1 mA dc ammeter (high resistance)
One 0–500 mA dc ammeter (high resistance)
Resistors—20 Ω, 1 kΩ, 10 kΩ, and 10 MΩ

Theory:

Electrical measuring instruments such as dc **voltmeters** and **ammeters** have **internal resistance**. When these instruments are connected to an electrical circuit, the instrument internal resistance is added to the circuit. In some circumstances this internal meter resistance will alter the original circuit enough to introduce a significant error in the voltage or current reading. This will usually happen when the resistance in the circuit is very high relative to the voltmeter resistance or very low relative to the ammeter resistance.

You can calculate the effect of the instrument internal resistance on the circuit by considering the instrument resistance as part of the circuit and including the meter resistance in your calculations. If the meter resistance (R_M) is unknown, you can determine its value by representing it in your circuit as an unknown resistance (R_M) and calculating the value of R_M from the known data in the circuit.

The circuit in Figure 11-1 will be used to demonstrate how adding a dc voltmeter to a circuit can change the circuit and introduce voltage measurement errors. The circuit in Figure 11-2 will be used to demonstrate how adding a dc ammeter to a circuit can change the circuit and introduce current measurement errors.

Figure 11-1 Measuring Voltage

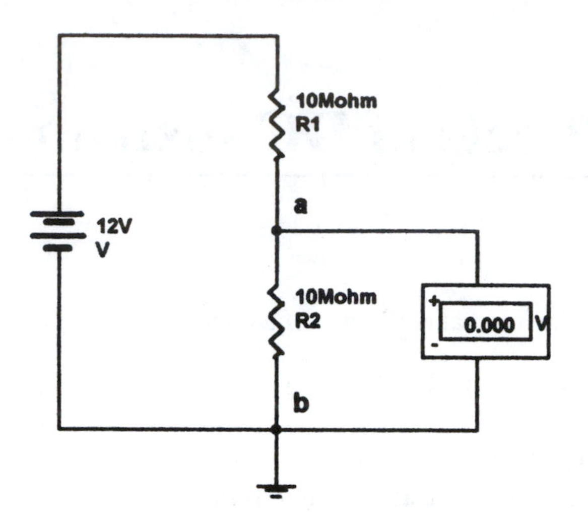

Figure 11-2 Measuring Current

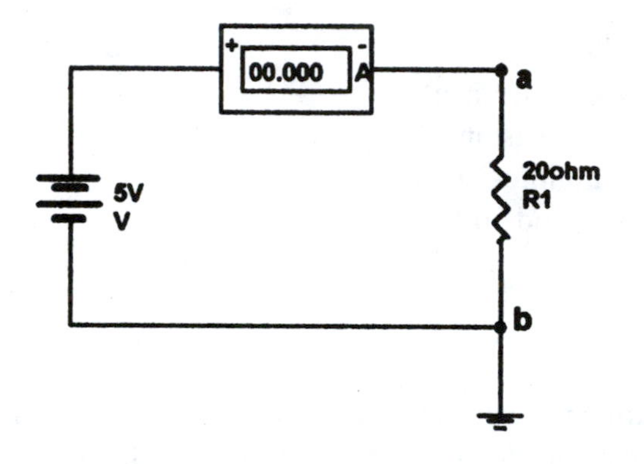

Procedure:

Step 1. Pull down the File menu and open FIG11-1. Click the On-Off switch to run the simulation. Record the voltage (V_{ab}) reading on the voltmeter.

V_{ab} = __________

Step 2. Calculate the expected value of voltage V_{ab} based on the component values in Figure 11-1, without the voltmeter present.

Question: How did the expected value of V_{ab} calculated in Step 2 compare with the value measured in Step 1? **Explain your answer.**

Step 3. Based on the value of V_{ab} measured in Step 1, calculate the internal resistance of the voltmeter (R_V).

Step 4. Change the value of R_1 and R_2 to 1 kΩ. Click the On-Off switch to run the simulation again. Record the voltage (V_{ab}) reading on the voltmeter.

V_{ab} = ___________

Step 5. Calculate the expected voltage V_{ab} based on the new values for R_1 and R_2 in Figure 11-1, without the voltmeter present.

Question: How did the expected value of V_{ab} calculated in Step 5 compare with the value measured in Step 4? **Explain your answer.**

Step 6. Pull down the File menu and open FIG11-2. Click the On-Off switch to run the simulation. Record the circuit current (I_{ab}) reading on the ammeter.

I_{ab} = __________

Step 7. Calculate the expected circuit current (I_{ab}) based on the component values in Figure 11-2, without the ammeter present.

Question: How did the expected value of I_{ab} calculated in Step 7 compare with the value measured in Step 6? **Explain your answer.**

Step 8. Based on the value of I_{ab} measured in Step 6, calculate the internal resistance of the ammeter (R_A).

Step 9. Change the value of R_1 to 10 kΩ. Click the On-Off switch to run the simulation again. Record the circuit current (I_{ab}) reading on the ammeter.

I_{ab} = __________

Step 10. Calculate the expected circuit current (I_{ab}) based on the new value of R_1 in Figure 11-2, without the ammeter present.

Question: How did the expected value of I_{ab} calculated in Step 10 compare with the value measured in Step 9? **Explain your answer.**

Troubleshooting Problems

1. Pull down the File menu and open FIG11-3. Click the On-Off switch to run the simulation. Is the voltmeter reading the correct reading? If not, why not?

2. Pull down the File menu and open FIG11-4. Click the On-Off switch to run the simulation. Is the voltmeter reading the correct reading? If not, why not?

3. Pull down the File menu and open FIG 11-5. Click the On-Off switch to run the simulation. Is the ammeter reading the correct reading? If not, why not?

EXPERIMENT

12

Name________________________

Date________________________

The Oscilloscope and Function Generator

Objectives:

1. Learn how to use the oscilloscope.
2. Learn how to use the function generator.

Theory:

The Oscilloscope

The **oscilloscope** is one of the most versatile and widely used electronic instruments. It displays a curve plot of a **time-varying voltage** on the oscilloscope screen, with the vertical axis representing voltage and the horizontal axis representing time. The oscilloscope provided with Electronics Workbench Multisim, shown in Figure 12-1, is a **dual-trace oscilloscope** that looks and acts like a real oscilloscope. A dual-trace oscilloscope allows the user to display and compare two time-varying voltages at one time.

By double clicking the left mouse button with the arrow on the oscilloscope, you can expand the oscilloscope to full screen display to adjust the controls and see the voltage curve plot on the oscilloscope screen during or after a circuit simulation run. If you wish to examine a curve plot in detail while the simulation is still running, you can **pause** the simulation by clicking the **pause button** next to the On-Off switch. Click the pause button again to **resume** the simulation. You can automatically show a single screen display by selecting **Sing** and **A** or **B** in the **Trigger section** on the oscilloscope. Then run the simulation to display the time-varying voltage on the oscilloscope screen after the trigger level is reached. Accurate voltage and time readings can be displayed by dragging the red and blue **cursors** to the points on the curve plot that you want to measure. The boxes below the screen display show the voltage and time for each cursor location and the difference between the readings for each cursor. The cursors are not displayed until a simulation occurs. The oscilloscope can be brought back to normal size by clicking the x in the upper right corner on the expanded oscilloscope.

The controls on the oscilloscope are as follows:

1. The **Time base scale** control adjusts the time scale on the horizontal axis in time per division when **Y/T** is selected. When **B/A** is selected, the horizontal axis no longer represents time. The horizontal axis now represents the voltage on the Channel A input and the vertical axis represents the voltage on the Channel B input. When **A/B** is selected, the horizontal axis represents the voltage on the Channel B input and the vertical axis represents the voltage on the Channel A input. The **X-pos** control determines the horizontal position where the curve plot begins.

2. The **Channel A scale** control adjusts the volts per division on the vertical axis for the Channel A input. The **Y-pos** control determines the vertical position of the Channel A curve plot relative to the horizontal axis. Selecting **AC** places a capacitance between the Channel A vertical input and the circuit test point. Selecting **0** connects the Channel A vertical input to ground.

3. The **Channel B scale** control adjusts the volts per division on the vertical axis for the Channel B input. The **Y-pos** control determines the vertical position of the Channel B curve plot relative to the horizontal axis. Selecting **AC** places a capacitance between the Channel B vertical input and the circuit test point. Selecting **0** connects the Channel B vertical input to ground.

4. The **Trigger** settings control the conditions under which a curve plot is triggered (begins to display). Triggering can be **internal** (based on one of the input signals) or **external** (based on a signal applied to the oscilloscope external trigger input). With **internal** triggering, you must select **Auto** (Automatic), or **Sing** (Single) and **A** or **B**, or **Nor** (Normal) and **A** or **B**. If **Sing** is selected, the oscilloscope will trigger one screen display after the trigger level is reached. If **Nor** is selected, the display will refresh every time the trigger level is reached. If **A** is selected, the curve plot will be triggered by the Channel A input signal. If **B** is selected, the curve plot will be triggered by the Channel B input signal. If you expect a flat input waveshape or you want the curve plot displayed as soon as possible, **Auto** should be selected. The display can be set to start on the positive or negative slope of the input curve plot by selecting the appropriate **Edge** selection. The trigger **Level** control determines the voltage level, in divisions on the vertical axis, of the input signal waveform that will trigger the display.

The Function Generator

A **function generator** is a **voltage source** that supplies different **time-varying voltage waveforms.** The output waveshape, frequency, amplitude, duty cycle, and dc offset can be easily changed. The function generator provided with Electronics Workbench Multisim, shown in Figure 12-1, can supply **sine wave, square wave**, and **triangular wave** time-varying voltage waveshapes and looks and acts like a real function generator. It has three voltage output terminals. The COM (common) terminal provides the reference level. To reference an output voltage from ground, connect the COM terminal to a ground symbol. The **positive terminal** (+) provides an output waveshape that is positive with reference to the COM terminal and the **negative terminal** (–) provides an output waveshape that is negative (inverted) with reference to the COM terminal.

The controls on the function generator are as follows:

1. You can select a **waveshape** by clicking the appropriate waveshape on the top of the function generator.

2. The **Frequency** control allows you to adjust the frequency of the output voltage curve plot between 1 Hz and 999 MHz. You can click the up or down arrow to adjust the frequency or you can click the frequency box and type the frequency desired.

3. The **Duty Cycle** control allows you to adjust the duty cycle of the output voltage curve plot between 1% and 99%. The default selection is 50. The duty cycle is the ratio of the time period when the output is high to the total time period of the waveshape. Duty cycle applies only to a

square wave or triangular wave output. A sine wave is not affected by the duty cycle setting. The duty cycle is adjusted in the same way that the frequency is adjusted.

4. The **Amplitude** control allows you to adjust the amplitude of the output voltage curve plot, measured from the reference level (common) to the peak level. The peak-to-peak amplitude will be twice the amplitude setting. The amplitude is adjusted in the same way that the frequency is adjusted.

5. The **Offset** control adjusts the dc level of the voltage curve plot generated by the function generator. An offset of 0 positions the voltage curve plot along the x-axis with an equal positive and negative amplitude. A positive offset raises the voltage curve plot above the x-axis and a negative offset lowers the voltage curve plot below the x-axis. The offset is adjusted in the same way that the frequency is adjusted.

Because the oscilloscope and function generator will be used extensively in the remaining experiments in this manual, a major purpose of this lab is to provide a tutorial on how to use them to generate and measure time-varying voltage functions. Consult the Electronics Workbench User Guide for more details on how to use these instruments. The knowledge gained in using these instruments will help you learn how to use a real oscilloscope and a real function generator. **If you are using this manual in a real laboratory environment, you should review the instrument manuals or see the instructor about the instruments you will be using before performing the following procedure.**

Figure 12-1 Function Generator and Oscilloscope

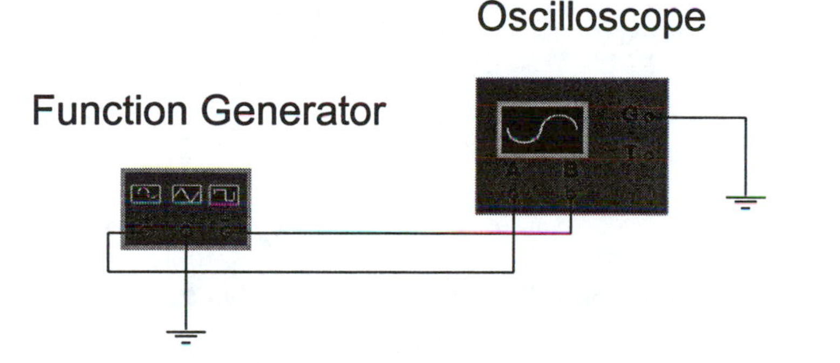

Procedure:

Step 1. Pull down the File menu and open FIG 12-1. Bring down the oscilloscope enlargement and make sure that the following settings are selected: Time base (Scale = 200 μs/Div, Xpos = 0, Y/T), CH A (Scale = 5 V/Div, Ypos = 0, DC), CH B (Scale = 5 V/Div, YPos = 0, DC), Trigger (Pos edge, Level = 0, Sing, A). Bring down the function generator enlargement and make sure that the following settings are selected: *Sine Wave*, Freq = 1 kHz, Ampl = 10 V, Offset = 0 V. Click the On-Off switch to run the simulation. Notice that you are plotting a sine wave function (red curve plot from the Channel A input) and a negative (inverted) sine wave function (blue curve plot from the Channel B input). Notice that the positive output of the function generator is feeding the oscilloscope Channel A input and the negative (inverted) output of the function generator is feeding the oscillocope Channel B input.

Step 2. Select the "0" on the oscilloscope Channel B input.

Question: What change occurred on the oscilloscope? **Explain.**

Step 3. Change the oscilloscope Channel A input to 10 V/Div.

Question: What change occurred on the oscilloscope Channel A curve plot? **Explain.**

Step 4. Change the oscilloscope TIME BASE to 100 μs/Div.

Question: What change occurred on the oscilloscope Channel A curve plot? **Explain.**

Step 5. Change the oscilloscope time base X-POS to 1.0.

Question: What change occurred on the oscilloscope Channel A curve plot? **Explain.**

Step 6. Change the oscilloscope trigger EDGE selection to negative edge triggering. Run the simulation again.

Question: What change occurred on the oscilloscope Channel A curve plot? **Explain.**

Step 7. Change the oscilloscope Channel A input back to 5 V/Div, change the trigger EDGE back to positive edge triggering, and change the trigger LEVEL to 5 V. Run the simulation again.

Question: What change occurred on the oscilloscope Channel A curve plot? **Explain.**

Step 8. Change the oscilloscope Channel A Y-POS to 1.0.

Question: What change occurred on the oscilloscope Channel A curve plot? **Explain.**

Step 9. Return the oscilloscope trigger LEVEL back to 0 V. Return the oscilloscope TIME BASE back to 200 μs/Div and the X-POS back to 0. Change the Channel A Y-POS back to 0 and select DC on the Channel B input. Select B/A in the Time base and run the simulation again.

Question: What change occurred on the oscilloscope curve plot? **Explain.**

Step 10. Return the oscilloscope time base back to Y/T and select "0" on the Channel B input again. Select the triangular waveshape on the function generator and run the simulation again.

Question: What change occurred on the oscilloscope Channel A curve plot? **Explain.**

Step 11. Select the square wave on the function generator and run the simulation again.

Question: What change occurred on the oscilloscope Channel A curve plot? **Explain.**

Step 12. Change the AMPLITUDE on the function generator to 5 V and run the simulation again.

Question: What change occurred on the oscilloscope Channel A curve plot? **Explain.**

Step 13. Change the FREQUENCY on the function generator to 2 kHz and run the simulation again.

Question: What change occurred on the oscilloscope Channel A curve plot? **Explain.**

Step 14. Change the OFFSET on the function generator to 2 V and run the simulation again.

Question: What change occurred on the oscilloscope Channel A curve plot? **Explain.**

Step 15. Change the DUTY CYCLE on the function generator to 25%. Run the simulation again.

Question: What change occurred on the oscilloscope Channel A curve plot? **Explain.**

Step 16. Place the crosshairs in position to measure the time period (T) of one cycle of the waveshape.

Question: What was the time period (T) of one cycle of the waveshape?

Step 17. Change the oscilloscope Trigger to Nor (Normal) and run the simulation.

Question: What change occurred on the oscilloscope? **Explain.**

Step 18. Stop the simulation. Change the oscilloscope Channel A input to AC and change the function generator DUTY CYCLE back to 50%. Run the simulation again.

Question: What change occurred on the oscilloscope? **Explain.**

PART

III

Capacitance and Inductance

The experiments in Part III involve the analysis of circuits that include capacitors and inductors. You will study series and parallel capacitors, series and parallel inductors, R-C circuits, R-L circuits, and R-L-C circuits. You will study the meaning of time constant as it applies to R-C and R-L circuits, and calculate and measure the time constant of an R-C circuit and an R-L circuit. You will study damping factor and resonant frequency as it applies to R-L-C circuits, and calculate and measure the damping factor and resonant frequency of an R-L-C circuit.

The circuits for the experiments in Part III can be found on the enclosed disk in the PART3 subdirectory.

EXPERIMENT

13

Name________________________________

Date________________________________

Capacitance—Series and Parallel Capacitors

Objectives:

1. Determine the capacitor current and voltage for a fully charged capacitor.
2. Determine the electrical charge stored in a charged capacitor.
3. Determine the electrical energy stored in a charged capacitor.
4. Determine the equivalent capacitance of capacitors connected in parallel.
5. Determine the equivalent capacitance of capacitors connected in series.
6. Determine the charge on each capacitor for charged capacitors in parallel.
7. Determine the voltage across each capacitor for charged capacitors in series.

Materials:

One 0–20 V dc voltage source
One dual-trace oscilloscope
One function generator
One 0–1 mA dc milliammeter
One 0–20 V dc voltmeter
One 1 kΩ resistor
Two 0.1 μF capacitors

Theory:

A **capacitor** is an electrical component that stores energy in the form of an electrical **charge** (Q). It consists of two conducting surfaces separated by an insulator, called a **dielectric**. The amount of charge (Q) that a capacitor stores is proportional to the applied voltage (V_C). The proportionality constant is the **capacitance** (C) of the capacitor in farads (F). Therefore,

$$Q = CV_C$$

The capacitance of a capacitor is one farad if it stores one coulomb of charge when the applied voltage across its terminals is one volt. A one farad capacitor would be extremely large. Therefore, practical capacitors range in size from microfarads ($1\ \mu F = 1 \times 10^{-6}$ F) to picofarads ($1\ pF = 1 \times 10^{-12}$ F). The capacitance of a capacitor is a measure of how much charge it will store per unit voltage applied across its terminals. The surface area of the two conducting surfaces, the thickness of the dielectric insulator between them, and the dielectric material determine the capacitance of a capacitor. The capacitance is proportional to the surface area of the conducting surfaces and inversely proportional to the thickness of the dielectric insulator between the conducting surfaces. The **electrical energy (W)** in joules stored in a capacitor can be found from the voltage across the capacitor (V_C) or the charge (Q) on the capacitor as follows:

$$W = \frac{1}{2}CV_C^2 = \frac{1}{2}\frac{Q^2}{C}$$

When a capacitor (C) is connected to a voltage source through a resistance (R), as shown in Figure 13-1, charge is transferred from one conducting surface of the capacitor to the other after the power is turned on, making one surface positive and the other surface negative. During this charging process, charges flow only through the voltage source (V) and the resistance (R). The rate of charge flow through the source and resistance is the charging current (I). There is no charge flow through the dielectric insulator separating the conducting surfaces of the capacitor. Therefore, the current flow is equal to the charging rate of the capacitor. When the capacitor approaches full charge, the charging rate of the capacitor approaches zero, causing the charging current (I) to approach zero. When the charging current (I) approaches zero, the voltage drop across resistor R will approach zero. This means that the voltage across the fully charged capacitor (V_C) will approach the supply voltage (V). This can be demonstrated in Figure 13-1 using Ohm's law and Kirchhoff's voltage law.

$$V = RI + V_C = 0 + V_C = V_C$$

When a capacitor charges through a resistance (R), a certain time interval is required for the capacitor to reach full charge. The rate at which the capacitor charges is determined by the value of the circuit resistance and capacitance. A charging capacitance takes a time interval of τ (one time constant) to reach 63% of its full charge voltage. In terms of the circuit capacitance (C) and resistance (R), τ can be found from

$$\tau = RC$$

where τ is the **time constant** in seconds, R is the resistance in ohms, and C is the capacitance in farads. **A capacitor is considered to be fully charged after a time interval of 5τ.**

If the voltage source is turned off, the charge (Q) will leak slowly through the dielectric insulator separating the conducting surfaces. The resistance of the dielectric insulator will determine how long the capacitor will retain the charge.

When capacitors are connected in **parallel**, the **equivalent capacitance (C_P)** is equal to the sum of the individual capacitances. Therefore, the equivalent capacitance in Figure 13-2 is

$$C_P = C_1 + C_2$$

The total charge transferred (Q_P) in a parallel capacitor circuit is equal to the sum of the charges on each capacitor. Therefore, in Figure 13-2

$$Q_P = Q_1 + Q_2$$

For the parallel capacitor circuit in Figure 13-2, one time constant (τ) can be found from

$$\tau = RC_P$$

When capacitors are connected in **series**, the **equivalent capacitance (C_S)** is found from the equation

$$\frac{1}{C_S} = \frac{1}{C_1} + \frac{1}{C_2} + \ldots + \frac{1}{C_n}$$

Therefore, the equivalent capacitance (C_S) in Figure 13-3 can be found from

$$\frac{1}{C_S} = \frac{1}{C_1} + \frac{1}{C_2}$$

The charge on each capacitor in a series capacitor circuit is equal to the total charge transferred (Q_S). Therefore, in Figure 13-3

$$Q_S = Q_1 = Q_2$$

For the series capacitor circuit in Figure 13-3, one time constant (τ) can be found from

$$\tau = RC_S$$

Figure 13-1 Current and Voltage in a Fully Charged Capacitor

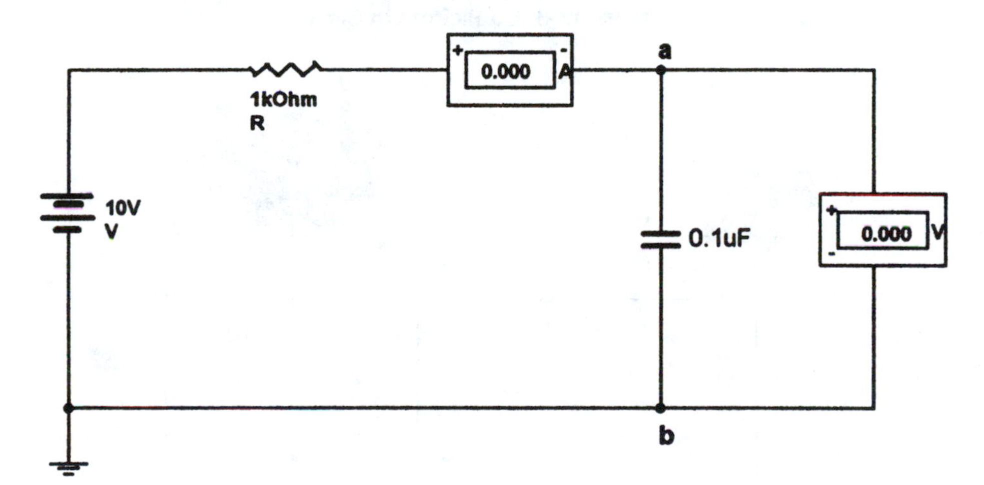

Figure 13-2 Capacitors in Parallel

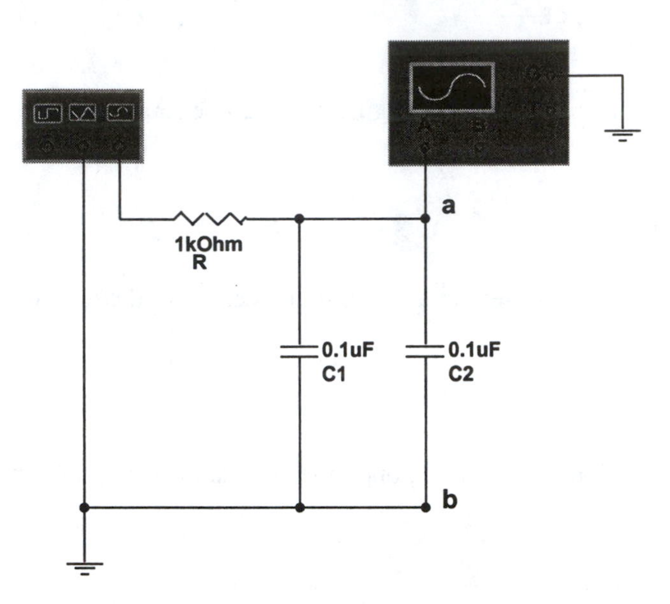

Figure 13-3 Capacitors in Series

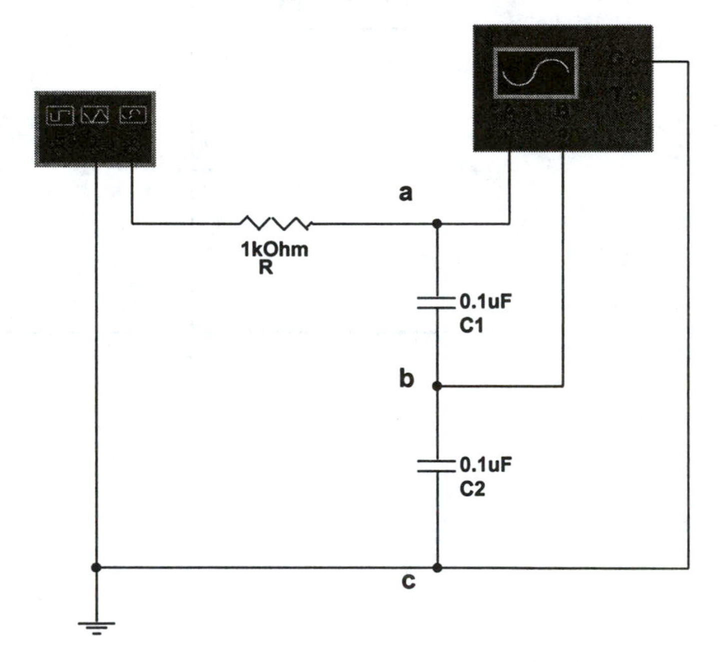

Procedure:

Step 1. Pull down the File menu and open FIG13-1. Click the On-Off switch to run the simulation. After the capacitor approaches full charge, record the capacitor voltage (V_{ab}) and the capacitor current (I_C).

V_{ab} = ______________ I_C = ______________

Question: What conclusion can you draw about the current in a fully charged capacitor? **Explain.**

Step 2. Based on the current (I_C) in a fully charged capacitor and the circuit component values in Figure 13-1, calculate the expected final voltage (V_{ab}) for a fully charged capacitor.

Questions: How did the calculated voltage (V_{ab}) in Step 2 compare with the measured voltage in Step 1?

What was the relationship between the voltage (V_{ab}) across the capacitor and the supply voltage (V) as the capacitor approached full charge? **Explain.**

Does a fully charged capacitor appear as an open circuit or a short circuit? **Explain.**

Step 3. Calculate the electrical charge (Q) on the capacitor based on the capacitor voltage (V_{ab}), for the fully charged capacitor.

Step 4. Calculate the electrical energy (W) stored in the fully charged capacitor.

Step 5. Pull down the File menu and open FIG13-2. Bring down the oscilloscope enlargement and make sure that the following settings are selected: Time base (Scale = 200 μs/Div, Xpos = 0, Y/T), Ch A (Scale = 5 V/Div, Ypos = 0, DC), Trigger (Pos edge, Level = 1 μV, Sing, A). Bring down the function generator enlargement and make sure that the following settings are selected: *Square Wave*, Freq = 200 Hz, Duty Cycle = 50%, Ampl = 5 V, Offset = 5 V. Click the On-Off switch to run the simulation. Notice that you are plotting the voltage across the charging capacitors (V_{ab}). Record the time it takes to reach 63% of the final voltage. This is the time constant (τ) for the R-C circuit. Also record the final steady-state voltage across the capacitors (V_{ab}) after the capacitors are fully charged.

τ = ______________ V_{ab} = ______________

Step 6. Based on the time constant (τ) measured in Step 5, calculate the value of the equivalent capacitance (C_P) for the parallel capacitors (C_1 and C_2).

Step 7. Based on the values of capacitors C_1 and C_2, calculate the equivalent capacitance (C_P) of the parallel capacitors in Figure 13-2.

Question: How did your measured value for C_P in Steps 5 and 6 compare with the calculated value in Step 7?

Step 8. Calculate the charge on each capacitor (Q_1 and Q_2) and the total charge (Q_P) on the parallel capacitors after the capacitors are fully charged (after five time constants).

Step 9. Pull down the File menu and open FIG13-3. Bring down the oscilloscope enlargement and make sure that the following settings are selected: Time base (Scale = 50 μs/Div, Xpos = 0, Y/T), Ch A (Scale = 5 V/Div, Ypos = 0, DC), Ch B (Scale = 5 V/Div, Ypos = 0, DC), Trigger (Pos edge, Level = 1 μV, Sing, A). Bring down the function generator enlargement and make sure that the following settings are selected: *Square Wave*, Freq = 200 Hz, Duty Cycle, = 50%, Ampl = 5 V, Offset = 5 V. Click the On-Off switch to run the simulation. Notice that you are plotting voltage V_{ac} across the series capacitors C_1 and C_2 (red curve plot) and voltage V_{bc} across capacitor C_2 (blue curve plot). Record the time it takes voltage V_{ac} to reach 63% of the final voltage. This is the time constant (τ) for the series capacitors. Also record the final steady-state voltage across the series capacitors (V_{ac}) and the final steady-state voltage across capacitor C_2 (V_{bc}) after the capacitors are fully charged.

τ = _______________ V_{ac} = _______________ V_{bc} = _______________

Step 10. Based on the time constant (τ) measured in Step 9, calculate the value of the equivalent capacitance (C_S) for the series capacitors (C_1 and C_2).

Step 11. Based on the values of capacitors C_1 and C_2, calculate the equivalent capacitance (C_S) of the series capacitors in Figure 13-3.

Question: How did your measured value for C_S in Steps 9 and 10 compare with the calculated value in Step 11?

Step 12. Based on the value of C_S and V_{ac}, calculate the charge (Q_S) on the series capacitors after the capacitors are fully charged.

Step 13. Based on the charge on the fully charged series capacitors, calculate the voltage across capacitor C_2.

Question: How did the measured voltage across capacitor C_2 (V_{bc}) in Step 9 compare with the calculated value in Step 13?

Step 14. Based on the voltage across series capacitors C_1 and C_2 (V_{ac}) and the voltage across capacitor C_2 (V_{bc}), calculate the voltage across capacitor C_1 (V_{ab}) after the capacitors are fully charged.

Troubleshooting Problems

1. Pull down the File menu and open FIG13-4. Click the On-Off switch to run the simulation. Based on the capacitor voltage and current, what is wrong with capacitor C?

2. Pull down the File menu and open FIG13-5. Click the On-Off switch to run the simulation. Measure the time constant (the time it takes for the voltage across the parallel capacitors to reach 63% of the final voltage). Based on the time constant (τ), which capacitor is open, C_1 or C_2?

3. Pull down the File menu and open FIG13-6. Click the On-Off switch to run the simulation. Measure the time constant (the time it takes for the voltage across the series capacitors to reach 63% of the final voltage). Based on the time constant (τ), which capacitor is shorted, C_1 or C_2?

EXPERIMENT

Name________________________________

Date________________________________

Charging and Discharging Capacitors

Objectives:

1. Determine the curve plot for the voltage across a charging capacitor as a function of time.
2. Determine the curve plot for the voltage across a discharging capacitor as a function of time.
3. Determine the curve plot for the current in a charging capacitor as a function of time.
4. Determine the curve plot for the current in a discharging capacitor as a function of time.
5. Measure the time constant of an R-C circuit and compare your measured value with the calculated value.
6. Determine the effect of changing the value of R and C on the time constant of an R-C circuit.

Materials:

One dual-trace oscilloscope
One function generator
Capacitors–0.1 μF, 0.2 μF
Resistors—1 kΩ, 2 kΩ

Theory:

Review the Theory section of Experiment 13 before attempting Experiment 14.

In Figure 14-1, when the function generator voltage switches from 0 V to 5 V, the charging voltage (v_C) across the capacitor will increase over a period of time determined by the **time constant** (**τ**) until the capacitor becomes fully charged. When the function generator voltage switches from 5 V to 0 V, the capacitor will discharge through resistor R and the capacitor voltage will decrease over a period of time determined by the time constant (τ), until it reaches zero. The time constant for the charging and discharging capacitor should be the same. The time constant can be measured from the capacitor voltage curve plot by determining the time required for the charging capacitor voltage (v_C) to rise to 63% of the final voltage during the charging process, or fall by 63% during the discharging process. The time constant can also be measured from the capacitor voltage curve plot by determining the time it would take for the voltage to rise to its final value, or fall to its final value, if it were to continue to rise, or fall, at its initial charging or discharging rate for the whole time interval.

In Figure 14-2, when the function generator voltage switches from 0 V to 5 V, the charging capacitor current (i_C) will start at a high positive value and decrease over a period of time determined by the time constant (τ), until the current reaches zero (capacitor is fully charged). When the function generator voltage switches from 5 V to 0 V, the capacitor current (i_C) will start at a high negative value and decrease over a period of time determined by the time constant (τ) as the capacitor discharges through resistor R, until it reaches zero. The time constant can be measured from the capacitor current curve plot by determining the time required for the capacitor current (i_C) to fall by 63%. The time constant can also be measured from the capacitor current curve plot by determining the time it would take for the current to fall to its final value if it continued to fall at its initial rate for the whole time interval. The charging or discharging capacitor current (i_C) in Figure 14-2 can be determined by determining the current in resistor R. The current in resistor R (i_R) can be calculated by dividing the voltage across the resistor (v_R) by the resistance (R). Therefore,

$$i_C = i_R = \frac{v_R}{R}$$

Figure 14-1 Charging and Discharging Capacitor Voltage

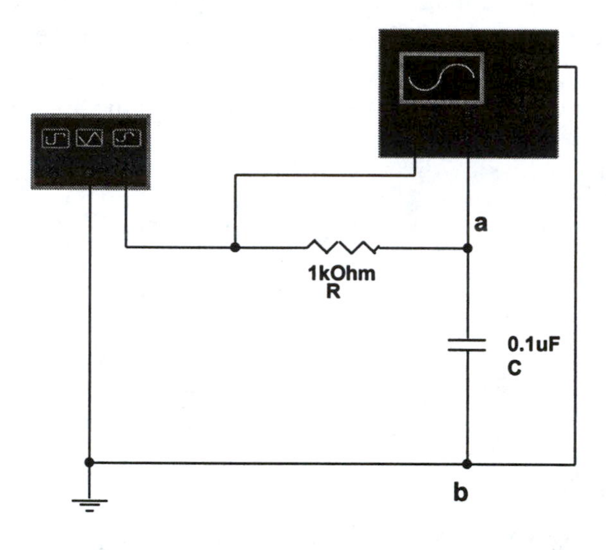

Figure 14-2 Charging and Discharging Capacitor Current

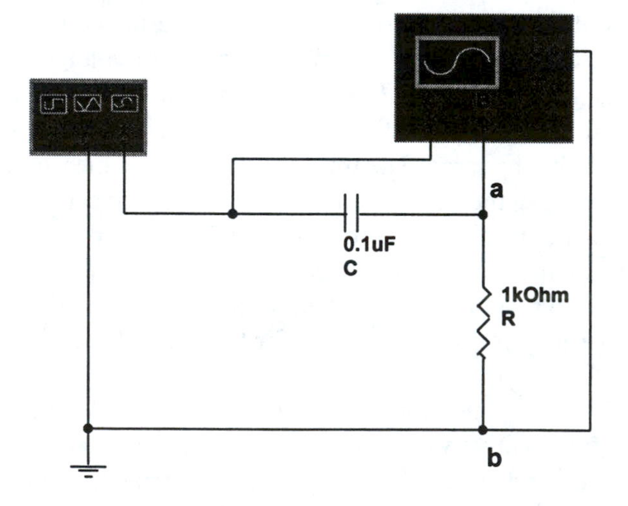

Procedure:

Step 1. Pull down the File menu and open FIG14-1. Bring down the oscilloscope enlargement and make sure that the following settings are selected: Time base (Scale = 100 μs/Div, Xpos = 0, Y/T), Ch A (Scale = 2 V/Div, Ypos = 0, DC), Ch B (Scale = 2 V/Div, Ypos = 0, DC), Trigger (Pos edge, Level = 1 μV, Sing, A). Bring down the function generator enlargement and make sure that the following settings are selected: *Square Wave*, Freq = 1 kHz, Duty Cycle = 50%, Ampl = 2.5 V, Offset 2.5 V. Click the On-Off switch to run the simulation. The red curve plot on the oscilloscope screen is the square wave output of the function generator. The generator output is switching between +5 V and 0 V, simulating the switching of a dc voltage source between +5 V and a short circuit. When the generator voltage switches from 0 V to +5 V, the capacitor will charge through resistor R. When the generator voltage switches from +5 V to 0 V (shorted to ground), the capacitor will discharge through resistor R. The blue curve is plotting the voltage across the capacitor (V_{ab}) as a function of time. Draw the curve plot of the capacitor voltage (V_{ab}) in the space provided. *Note on the drawing the part of the curve plot that represents the capacitor charging voltage and the part that represents the capacitor discharging voltage.*

Questions: What is the final voltage across the capacitor (V_{ab}) when it is fully charged? How does it compare with the supply voltage? **Explain.**

What is the voltage across the capacitor (V_{ab}) when it is fully discharged? Is it what you expected? **Explain.**

Step 2. Measure the time constant (τ) for the R-C circuit from the curve plot and record your answer.

τ = __________

Step 3. Based on the value of R and C in Figure 14-1, calculate the expected time constant (τ) for the R-C circuit.

Questions: How did your calculated value for the time constant (τ) compare with the value measured in Step 2?

How many time constants did it take for the capacitor to reach full charge? Full discharge?

Step 4. Pull down the File menu and open FIG14-2. Bring down the oscilloscope enlargement and make sure that the following settings are selected: Time base (Scale = 100 μs/Div, Xpos = 0, Y/T), Ch A (Scale = 2 V/Div, Ypos = 0, DC), Ch B (Scale = 2 V/Div, Ypos = 0, DC), Trigger (Pos edge, Level = 1 μV, Sing, A). Bring down the function generator enlargement and make sure that the following settings are selected: *Square Wave*, Freq = 1 kHz, Duty Cycle = 50%, Ampl = 2.5 V, Offset 2.5 V. Click the On-Off switch to run the simulation. The red curve plot on the oscilloscope screen is the square wave output of the function generator. The generator output is switching between +5 V and 0 V, simulating the switching of a dc voltage source between +5 V and a short circuit. When the generator voltage switches from 0 V to +5 V, the capacitor will charge through resistor R. When the generator voltage switches from +5 V to 0 V (shorted to ground), the capacitor will discharge through resistor R. The blue curve is plotting the voltage across resistor R (which is proportional to the capacitor current) as a function of time. Draw the curve plot of the resistor voltage (capacitor current) in the space provided on the following page. *Note on the drawing the part of the curve plot that represents the charging capacitor and the part that represents the discharging capacitor.*

Step 5. Based on the value of resistor R and the curve plot voltage readings, calculate the value of the capacitor current (i_C) when charging begins.

Step 6. Based on the value of resistor R and the curve plot voltage readings, calculate the value of the capacitor current (i_C) when discharging begins.

Questions: What is the capacitor current when the capacitor is fully charged?

What is the capacitor current when the capacitor is fully discharged?

Why was the capacitor discharging current negative? **Explain.**

Step 7. Measure the time constant (τ) for the R-C circuit from the curve plot and record your answer.

$\tau =$ __________

Questions: How did the measured time constant in Step 7 compare with the value calculated in Step 3?

Why are the capacitor current and capacitor voltage time constants the same value? **Explain.**

Step 8. Change R to 2 kΩ. Click the On-Off switch to run the simulation again. Measure the new time constant (τ) from the curve plot and record your answer.

τ = __________

Step 9. Based on the new value of R, calculate the new time constant (τ) for the R-C circuit in Figure 14-2.

Question: What effect did changing the value of R have on the time constant? **Explain.**

Step 10. Change C to 0.2 μF. Change the frequency of the function generator to 500 Hz. Click the On-Off switch to run the simulation again. Measure the new time constant (τ) from the curve plot and record your answer.

τ = __________

Step 11. Based on the new values of R and C, calculate the new time constant (τ) for the R-C circuit in Figure 14-2.

Question: What effect did changing the value of C have on the time constant? **Explain.**

Troubleshooting Problems

1. Pull down the File menu and open FIG14-3. Click the On-Off switch to run the simulation. Measure the time constant from the capacitor voltage curve plot. Based on the measured time constant, determine the value of R.

   R = __________

2. Pull down the File menu and open FIG14-4. Click the On-Off switch to run the simulation. Measure the time constant from the capacitor voltage curve plot. Based on the measured time constant, determine the value of C.

   C = __________

EXPERIMENT

15

Name________________________________

Date________________________________

Inductance—Series and Parallel Inductors

Objectives:

1. Determine inductor current and voltage after the inductor current has reached a final steady-state value.
2. Determine the electrical energy stored in the magnetic field of an inductor after the inductor current has reached a final steady-state value.
3. Determine the equivalent inductance of inductors connected in series.
4. Determine the equivalent inductance of inductors connected in parallel.

Materials:

One 0–20 V dc voltage source
One 0–10 mA dc milliammeter
One 0–10 V dc voltmeter
One function generator
One dual-trace oscilloscope
Two 100 mH inductors
Resistors—250 Ω, 1 kΩ

Theory:

When a length of wire is wound into a coil on a cylindrical core, it forms an **inductor**. When there is an electrical current through the coil, a magnetic field proportional to the magnitude of the current is created, surrounding the coil. When the current magnitude changes, the magnetic field strength changes. This changing field cutting through the coil causes an **induced voltage** (V_L) across the coil terminals that is proportional to the rate of change of the magnetic field. This is **Faraday's law**. Because the magnetic field strength is proportional to the magnitude of the changing current, the magnitude of the induced voltage across the coil is proportional to the rate of change of the current. The proportionality constant is called the **inductance** (L) of the coil in henries (H). This can be represented mathematically as

$$V_L = L\frac{di}{dt}$$

where $\frac{di}{dt}$ is the rate of change of the current in amperes per second.
The inductance of a coil is one henry if the induced voltage created by its changing current is one volt when the current changes at a rate of one ampere per second. In many practical applications, inductances are in millihenries (1 mH = 1×10^{-3} H) or microhenries (1 μH = 1×10^{-6} H). The

inductance of an inductor is a measure of how strong a magnetic field it will produce per unit of current through the coil, or a measure of the inductor's ability to produce an induced voltage across its terminals as a result of a changing current through the coil. The cross sectional area of the coil, the coil length, the number of turns of wire, and the core material determines the inductance of the coil. The inductance is proportional to the coil cross sectional area and the number of coil turns squared, and inversely proportional to the length of the coil. Inductors with a high inductance are usually wound on a steel core.

When the current and magnetic field are increasing, the induced voltage across the coil is a voltage drop in the direction of the current, causing it to act like a load. When the current and magnetic field are decreasing, the induced voltage across the coil is a voltage rise in the direction of the current, causing it to act like a source. Therefore, the induced voltage is always in a direction that opposes the change in the current and magnetic field. This is **Lenz's law**.

An inductor is an electrical component that stores energy in a magnetic field. The inductance (L) is a measure of how much energy it will store per unit of electrical current through the coil. The **electrical energy (W)** stored in an inductor magnetic field is proportional to the square of the electrical current (i_L) in the coil and can be found from:

$$W = \frac{1}{2} L {i_L}^2$$

Because the induced voltage across an inductor is in such a direction to oppose the change in the current (Lenz's law), the current cannot change instantaneously in an inductor. When an inductor (L) is connected to a voltage source through a resistor (R), as shown in Figure 15-1, the inductor current will build up until it reaches a final steady state (constant) value. The voltage across the inductor will approach zero as the inductor current approaches steady state (a constant value). The inductor voltage (V_{ab}) is proportional to the rate of change of the inductor current (di/dt), which is zero when the current is constant. This can be demonstrated with the following equation:

$$V_{ab} = L\frac{di}{dt} = L(0) = 0$$

This means that the inductor looks like a short circuit when the inductor current is at its final steady-state value, and the supply voltage will be across resistor R. Therefore, the final steady-state current in the inductor (I_L) can be found from Ohm's law and Kirchhoff's voltage law as follows:

$$V = I_L R + V_{ab} = I_L R + 0 = I_L R$$

$$I_L = \frac{V}{R}$$

In a real laboratory environment, the voltage across the inductor may not be exactly zero for a steady-state (constant) current because the **inductor has a small resistance (R_L) due to the resistance of the coil wire. This resistance should be added to the resistance of R in all of the equations when performing this experiment in a real laboratory environment.**

The time required for the magnetic field and the current to build up to its final steady-state (constant) value is determined by the circuit resistance and inductance. It takes a time interval of τ (one time constant) for the current to reach 63% of its final value. In terms of the circuit inductance (L) and resistance (R), one time constant can be found from

$$\tau = \frac{L}{R}$$

where τ is the **time constant** in seconds, R is the circuit resistance in ohms, and L is the inductance in henries. **An inductor current is considered to have reached its final value (steady state) after a time interval of 5τ.**

The **equivalent inductance** for inductors in **series** (L_S) is equal to the sum of the individual inductances. Therefore, the equivalent inductance in Figure 15-2 is

$$L_S = L_1 + L_2$$

For the series inductor circuit in Figure 15-2 the time constant is

$$\tau = \frac{L_S}{R}$$

The **equivalent inductance** for inductors in **parallel** (L_P) is found from the equation

$$\frac{1}{L_P} = \frac{1}{L_1} + \frac{1}{L_2} + \ldots + \frac{1}{L_n}$$

Therefore, the equivalent inductance in Figure 15-3 can be found from

$$\frac{1}{L_P} = \frac{1}{L_1} + \frac{1}{L_2}$$

For the parallel inductor circuit in Figure 15-3 the time constant is

$$\tau = \frac{L_P}{R}$$

Figure 15-1 Inductance—DC Analysis

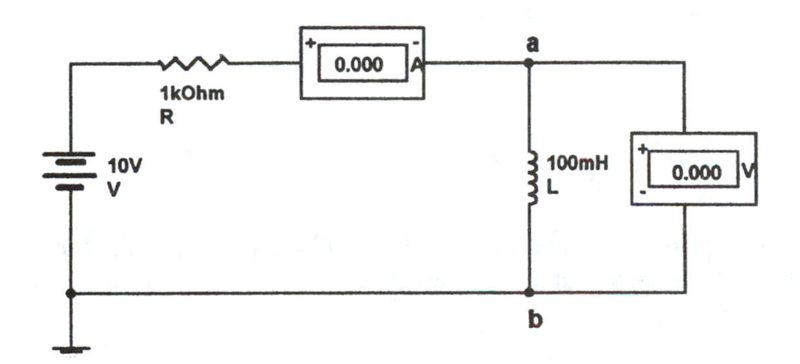

Figure 15-2 Inductors in Series

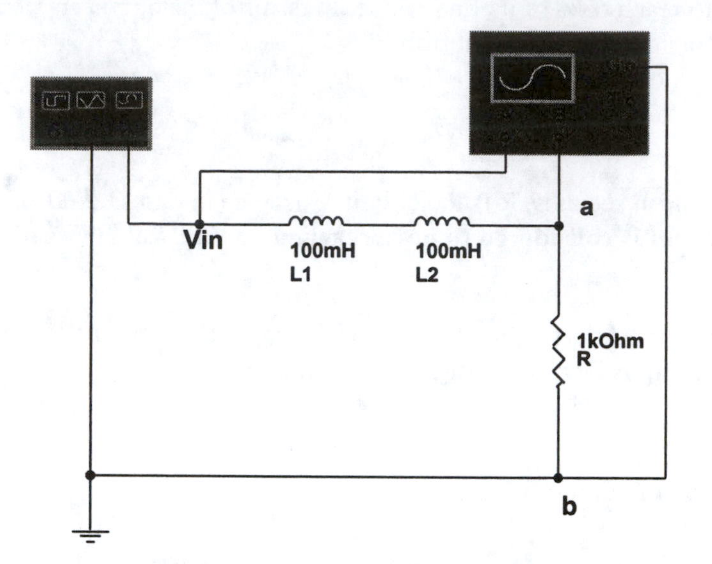

Figure 15-3 Inductors in Parallel

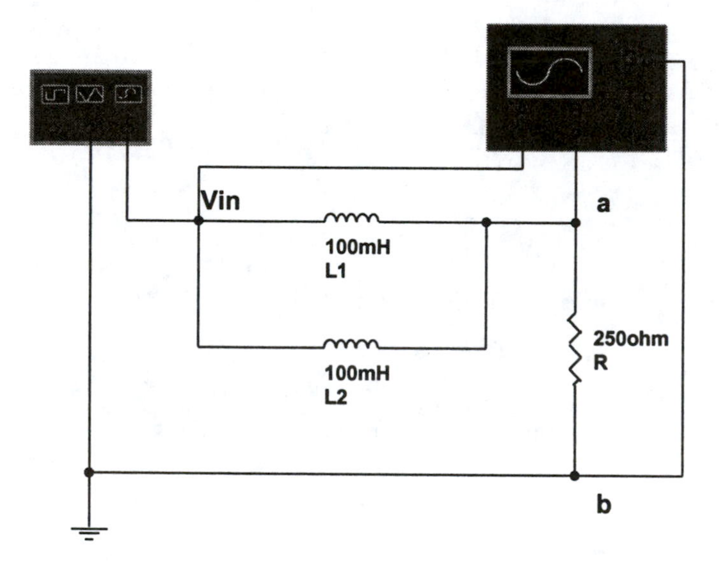

Procedure:

Step 1. Pull down the File menu and open FIG15-1. Click the On-Off switch to run the simulation. After steady state is reached, record the inductor voltage (V_{ab}) and the steady-state inductor current (I_L).

V_{ab} = __________ I_L = __________

Questions: What conclusion can you draw about the voltage across an inductor after the inductor current has reached steady state and is constant? **Explain.**

When the current in an inductor has reached steady state, does the inductor appear as a short circuit or an open circuit? **Explain.**

Step 2. Based on the circuit component values in Figure 15-1, calculate the final steady-state inductor current (I_L) and the final steady-state inductor voltage (V_{ab}).

Question: How did the calculated values for the steady-state inductor current and voltage compare with the values measured in Step 1?

Step 3. Based on the steady-state inductor current (I_L), calculate the electrical energy (W) stored in the inductor magnetic field.

Step 4. Pull down the File menu and open FIG15-2. Bring down the oscilloscope enlargement and make sure that the following settings are selected: Time base (Scale = 100 μs/Div, Xpos = 0, Y/T), Ch A (Scale = 5 V/Div, Ypos = 0, DC), Ch B (Scale = 5 V/Div, Ypos = 0, DC), Trigger (Pos edge, Level = 1 μV, Sing A). Bring down the function generator enlargement and make sure that the following settings are selected: *Square Wave*, Freq = 100 Hz, Duty Cycle = 90%, Ampl = 5 V, Offset = 5 V. Click the On-Off switch to run the simulation. You are looking at the curve plots for the voltage across resistor R (blue) and the input voltage (red). The voltage across R is proportional to the current in the inductors. Record the time it takes for the voltage across resistor R (inductor current) to reach 63% of the final value. This is one time constant (τ) for the R-L network. Also record the final steady-state voltage across resistor R (V_{ab}).

τ = __________ V_{ab} = __________

Step 5. Based on the time constant (τ) measured in Step 4, calculate the value of the equivalent inductance (L_S) for the series inductors (L_1 and L_2).

Step 6. Based on the values of inductors L_1 and L_2, calculate the equivalent inductance (L_S) of the series inductors in Figure 15-2.

Question: How did your measured value of L_S in Steps 4 and 5 compare with the calculated value in Step 6?

Step 7. Calculate the final steady-state inductor current (I_L) based on the final steady-state voltage V_{ab} across resistor R, measured in Step 4.

Step 8. Pull down the File menu and open FIG15-3. Bring down the oscilloscope enlargement and make sure that the following settings are selected: Time base (Scale = 100 μs/Div, Xpos = 0, Y/T), Ch A (Scale = 5 V/Div, Ypos = 0, DC), Ch B (Scale = 5 V/Div, Ypos = 0, DC), Trigger (Pos edge, Level = 1 μV, Sing A). Bring down the function generator enlargement and make sure that the following settings are selected: *Square Wave*, Freq = 100 Hz, Duty Cycle = 90%, Ampl = 5 V, Offset = 5 V. Click the On-Off switch to run the simulation. You are looking at the curve plots for the voltage across resistor R (blue) and the input voltage (red). The voltage across R is proportional to the current in the inductors. Record the time it takes for the voltage across resistor R (inductor current) to reach 63% of the final value. This is one time constant (τ) for the R-L network.

τ = __________

Step 9. Based on the time constant (τ) measured in Step 8, calculate the value of the equivalent inductance (L_P) for the parallel inductors (L_1 and L_2).

Step 10. Based on the values of inductors L_1 and L_2, calculate the equivalent inductance (L_P) of the parallel inductors in Figure 15-3.

Question: How did your measured value of L_P in Steps 8 and 9 compare with the calculated value in Step 10?

Troubleshooting Problems

1. Pull down the File menu and open FIG15-4. Click the On-Off switch to run the simulation. Based on the inductor voltage and current, what is wrong with inductor L?

2. Pull down the File menu and open FIG15-5. Click the On-Off switch to run the simulation. Measure the time constant (τ) for the R-L network. Based on the time constant, which inductor is shorted, L_1 or L_2?

3. Pull down the File menu and open FIG15-6. Click the On-Off switch to run the simulation. Measure the time constant (τ) for the R-L network. Based on the time constant, which inductor is open, L_1 or L_2?

EXPERIMENT

16

Name____________________

Date____________________

Transients in Inductors

Objectives:

1. Determine the curve plot for the current in an inductor as a function of time when the inductor current is increasing.
2. Determine the curve plot for the current in an inductor as a function of time when the inductor current is decreasing.
3. Determine the curve plot for the induced voltage across an inductor as a function of time when the inductor current is increasing.
4. Determine the curve plot for the induced voltage across an inductor as a function of time when the inductor current is decreasing.
5. Measure the time constant of an R-L circuit and compare your measured value with the calculated value.
6. Determine the effect of changing the value of R and L on the time constant of an R-L circuit.

Materials:

One dual-trace oscilloscope
One function generator
Inductors—100 mH, 200 mH
Resistors—1 kΩ, 2 kΩ

Theory:

Review the Theory section of Experiment 15 before attempting Experiment 16.

In Figure 16-1, when the function generator voltage switches from 0 V to 10 V, the inductor current (i_L) will increase over a period of time determined by the **time constant (τ)** until it reaches a steady-state (constant) value. When the function generator voltage switches from 10 V to 0 V, the inductor current will decrease through resistor R over a period of time determined by the time constant (τ), until it reaches zero. The time constant for the increasing and decreasing current should be the same. The time constant can be measured from the inductor current curve plot by determining the time required for the inductor current (i_L) to rise to 63% of the final value on the increasing current curve plot, or fall by 63% on the decreasing current curve plot. The time constant can also be measured from the inductor current curve plot by determining the time it would take for the current to rise to its final value, or fall to its final value, if it were to continue to rise, or fall, at its initial rate for the whole time interval. The inductor current (i_L) in Figure 16-1 can be determined by determining the current in resistor R. The current in resistor R (i_R) can be calculated by dividing the voltage across the resistor (v_R) by the resistance (R).

Therefore,

$$i_L = i_R = \frac{v_R}{R}$$

In Figure 16-2, the induced voltage across the inductor (v_L) is positive when the inductor current is increasing (di/dt is positive) and is negative when the inductor current is decreasing (di/dt is negative). The maximum positive induced voltage across the inductor occurs when the inductor current is just starting to increase. At this time the inductor current is zero, but it is rising at its fastest rate. When the current is zero, the voltage across resistor R is zero, leaving all of the supply voltage across the inductor. Therefore, when the function generator voltage goes from 0 V to 10 V, the induced voltage across the inductor (v_L) will start at 10 V and decrease over a period of time determined by the time constant (τ), until the voltage reaches zero (constant inductor current). The maximum negative induced voltage across the inductor occurs when the inductor current is just starting to decrease. At this time the inductor current is at its maximum value, but it is falling at its fastest rate. The inductor voltage is negative because the current is decreasing, causing di/dt to be negative $\left(v_L = L\frac{di}{dt}\right)$. Therefore, when the function generator voltage goes from 10 V to 0 V, the induced voltage across the inductor (v_L) will start at -10 V and decrease in magnitude over a period of time determined by the time constant (τ), until it reaches zero (magnetic field collapsed to zero). The time constant can be measured from the inductor voltage curve plot by determining the time required for the inductor voltage (v_L) to fall by 63%. The time constant can also be measured from the inductor voltage curve plot by determining the time it would take for the voltage to fall to its final value if it continued to fall at its initial rate for the whole time interval.

Figure 16-1 Transients in Inductors—Inductor Current

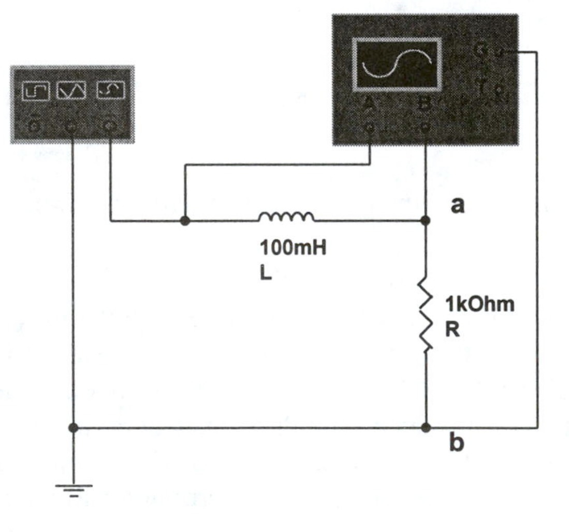

Figure 16-2 Transients in Inductors—Inductor Voltage

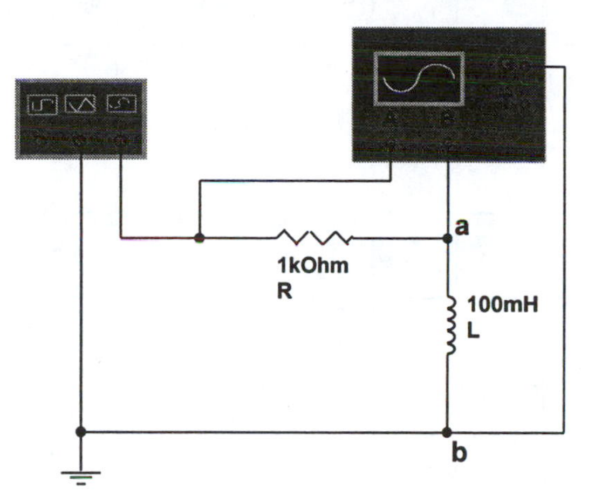

Procedure:

Step 1. Pull down the File menu and open FIG16-1. Bring down the oscilloscope enlargement and make sure that the following settings are selected: Time base (Scale = 100 μs/Div, Xpos = 0, Y/T), Ch A (Scale = 5 V/Div, Ypos = 0, DC), Ch B (Scale = 5 V/Div, Ypos = 0, DC), Trigger (Pos edge, Level = 1 μV, Sing, A). Bring down the function generator enlargement and make sure that the following settings are selected: *Square Wave*, Freq = 1 kHz, Duty Cycle = 50%, Ampl = 5 V, Offset = 5 V. Click the On-Off switch to run the simulation. The red curve plot on the oscilloscope screen is the square wave output of the function generator. The generator output is switching between +10 V and 0 V, simulating the switching of a dc voltage source between +10 V and a short circuit. When the generator voltage goes from 0 V to +10 V, the inductor current will increase until it reaches a maximum steady-state value. When the generator voltage goes from +10 V to 0 V (shorted to ground), the inductor current will decrease until it reaches zero. The blue curve is plotting the voltage across resistor R (which is proportional to the inductor current) as a function of time. Draw the curve plot of the resistor voltage (inductor current) in the space provided on the following page. *Note on the drawing the part of the curve plot that represents the increasing inductor current and the part that represents the decreasing inductor current.*

Step 2. Based on the value of resistor R and the curve plot voltage readings, calculate the final inductor current (I_L) when the inductor current has reached a maximum steady-state value.

Step 3. Based on the value of resistor R and the +10 V generator output voltage, calculate the expected value of the inductor current (I_L) when it reaches the maximum steady-state value.

Questions: How did the value of the maximum steady-state inductor current measured in Steps 1 and 2 compare with the value calculated in Step 3?

Predict the voltage across the inductor when the inductor current reaches the maximum steady-state value. **Explain.**

Step 4. Measure the time constant (τ) for the R-L circuit from the curve plot and record your answer.

τ = __________

Step 5. Based on the value of R and L in Figure 16-1, calculate the expected time constant (τ) for the R-L circuit.

Questions: How did your calculated value for the time constant (τ) compare with the value measured in Step 4?

How many time constants did it take for the inductor current to reach the maximum steady-state value?

Step 6. Pull down the File menu and open FIG16-2. Bring down the oscilloscope enlargement and make sure that the following settings are selected: Time base (Scale = 100 μs/Div, Xpos = 0, Y/T), Ch A (Scale = 5 V/Div, Ypos = 0, DC), Ch B (Scale = 5 V/Div, Ypos = 0, DC), Trigger (Pos edge, Level = 1 μV, Sing, A). Bring down the function generator enlargement and make sure that the following settings are selected: *Square Wave*, Freq = 1 kHz, Duty Cycle = 50%, Ampl = 5 V, Offset = 5 V. Click the On-Off switch to run the simulation. The red curve plot on the oscilloscope screen is the square wave output of the function generator. The generator output is switching between +10 V and 0 V, simulating the switching of a dc voltage source between +10 V and a short circuit. When the generator voltage goes from 0 V to +10 V, the inductor current will increase until it reaches a maximum steady-state value, causing the inductor voltage to decrease to zero when the inductor current has reached steady state. When the generator voltage goes from +10 V to 0 V (shorted to ground), the inductor current will decrease until it reaches zero, causing the inductor voltage to be negative and decrease in value as the current goes to zero. The blue curve is plotting the voltage across the inductor (V_{ab}) as a function of time. Draw the curve plot of the inductor voltage (V_{ab}) in the space provided on the following page. *Note on the drawing the part of the curve plot that represents the inductor voltage when the inductor current is increasing and the part that represents the inductor voltage when the inductor current is decreasing.*

Questions: What is the maximum inductor voltage when the inductor current is increasing? **Explain.**

What is the maximum inductor voltage when the inductor current is decreasing? Why is it negative? **Explain.**

What is the minimum inductor voltage (V_{ab})? **Explain.**

Step 7. Measure the time constant (τ) for the R-L circuit from the curve plot and record your answer.

$\tau =$ __________

Questions: How did the measured time constant in Step 7 compare with the value calculated in Step 5?

Why are the inductor current and inductor voltage time constants the same value? **Explain.**

Step 8. Change R to 2 kΩ. Click the On-Off switch to run the simulation again. Measure the new time constant (τ) from the curve plot and record your answer.

$\tau =$ __________

Step 9. Based on the new value of R, calculate the new time constant (τ) for the R-L circuit in Figure 16-2.

Question: What effect did changing the value of R have on the time constant? **Explain.**

Step 10. Change L to 200 mH. Click the On-Off switch to run the simulation again. Measure the new time constant (τ) from the curve plot and record your answer.

$\tau =$ __________

Step 11. Based on the new values of R and L, calculate the new time constant (τ) for the R-L circuit in Figure 16-2.

Question: What effect did changing the value of L have on the time constant? **Explain.**

Troubleshooting Problems

1. Pull down the File menu and open FIG16-3. Click the On-Off switch to run the simulation. Measure the time constant from the curve plot. Based on the measured time constant, determine the value of L.

   L = __________

Step 2. Pull down the File menu and open FIG16-4. Click the On-Off switch to run the simulation. Measure the time constant from the curve plot. Based on the measured time constant, determine the value of R.

   R = __________

EXPERIMENT

17

Name____________________

Date____________________

Transients in R-L-C Circuits

Objectives:

1. Determine the step response for the current in a series R-L-C circuit that is overdamped.
2. Determine the step response for the current in a series R-L-C circuit that is underdamped.
3. Determine the damping factor and the resonant frequency for a series R-L-C circuit.
4. Demonstrate how changing the value of R affects the damping factor of a series R-L-C circuit.
5. Demonstrate how changing the value of C affects the frequency of oscillation of an underdamped R-L-C circuit.

Materials:

One dual-trace oscilloscope
One function generator
One 10 mH inductor
Capacitors—0.05 μF, 0.1μF
Resistors—200 Ω, 400 Ω, 1 kΩ

Theory:

In this experiment, you will determine the **step response** for the current in the series R-L-C circuit shown in Figure 17-1. This means that you will plot the current when the input voltage steps from zero to a constant (steady-state) value. The step input voltage will be provided by a function generator that produces a square wave output with a time period that is much longer than the transient response of the R-L-C circuit.

The current in a series R-L-C circuit can have three possible step responses, depending on the relationship between the circuit **damping factor** (α) and **resonant frequency** (ω_o). They are overdamped, critically damped, or underdamped. The damping factor (α) for a series R-L-C circuit is dependent on the circuit resistance (R) and inductance (L). The damping factor, in sec^{-1}, for a series R-L-C circuit is calculated from

$$\alpha = \frac{R}{2L}$$

The resonant frequency (ω_o) for a series R-L-C circuit is dependent on the circuit inductance (L) and capacitance (C). The resonant frequency (ω_o), in radians/sec, for a series R-L-C circuit is calculated from

$$\omega_0 = \frac{1}{\sqrt{LC}}$$

When the damping factor (α) is equal to the resonant frequency (ω_o), the series R-L-C circuit is **critically damped**. When the damping factor (α) is greater than the resonant frequency (ω_o), the series R-L-C circuit is **overdamped**. When the damping factor (α) is less than the resonant frequency (ω_o), the series R-L-C circuit is **underdamped**.

The damping factor (α) determines approximately how long it takes for the current to damp out to a steady-state value (in this case zero). The current will damp out to steady state in approximately five time constants. One time constant (τ) can be found from

$$\tau = \frac{1}{\alpha}$$

The resonant frequency (ω_o) is the frequency of oscillation when the R-L-C circuit is underdamped with a damping factor (α) of zero, which would make it undamped. The **damped frequency of oscillation (ω_d)** of an underdamped R-L-C circuit is less than the resonant frequency (ω_o) and is calculated, in radians/sec, from

$$\omega_d = \sqrt{\omega_0^2 - \alpha^2}$$

The frequency of the underdamped oscillation can be determined from the curve plot by measuring the time period (T_d) for one oscillation cycle. The frequency in hertz (cycles per second) can be calculated from

$$f_d = \frac{1}{T_d}$$

The frequency in radians per second (ω_d) can be calculated from f_d using the equation

$$\omega_d = 2\pi f_d$$

In a real laboratory environment, the inductor has a small resistance (R_L). This resistance should be added to the resistance of R in all of the equations.

Figure 17-1 Series R-L-C Circuit

10mH
L
0.1uF
C
1kOhm
R

Procedure:

Step 1. Pull down the File menu and open FIG17-1. Bring down the oscilloscope enlargement and make sure that the following settings are selected: Time base (Scale = 50 μs/Div, Xpos = 0, Y/T), Ch A (Scale = 2 V/Div, Ypos = 0, DC), Ch B (Scale = 2 V/Div, Ypos = 0, DC), Trigger (Pos edge, Level = 1 μV, Sing, A). Bring down the function generator enlargement and make sure that the following settings are selected: *Square Wave*, Freq = 500 Hz, Duty Cycle = 50%, Ampl = 2.5 V, Offset = 2.5 V. Click the On-Off switch to run the simulation. The red curve plot on the oscilloscope screen is the output of the function generator, which is the applied step voltage to the series R-L-C circuit. The blue curve plot is the voltage across resistor R, which is proportional to the current in the R-L-C circuit. Draw the curve plot of the resistor voltage (circuit current) in the space provided.

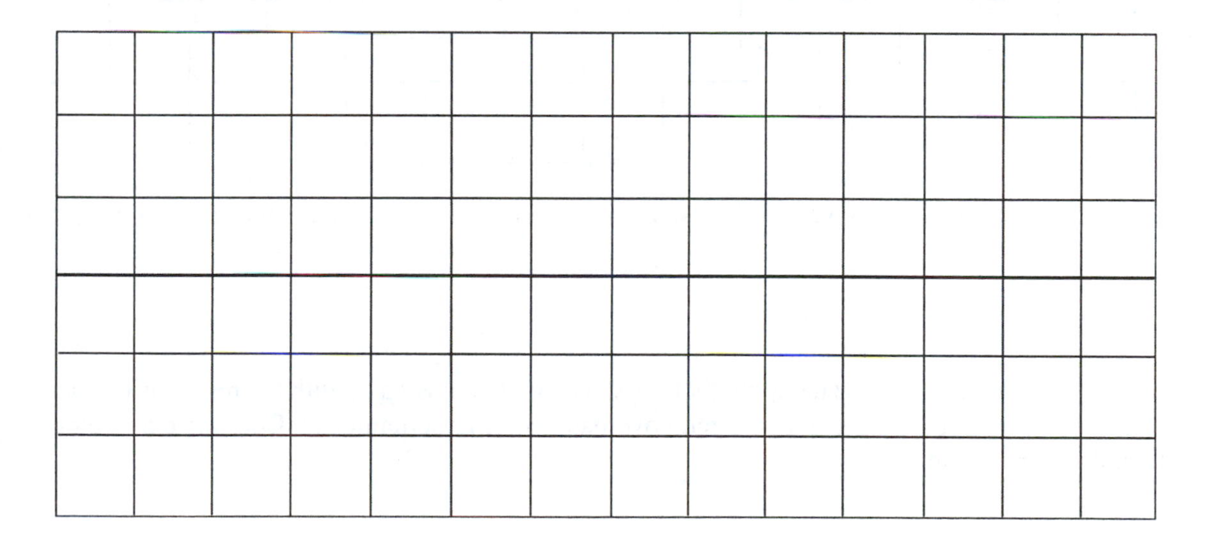

Step 2. Based on the R-L-C circuit component values in Figure 17-1, calculate the damping factor (α

Step 3. Based on the R-L-C circuit component values in Figure 17-1, calculate the resonant frequency (ω_0), in radians per second.

Question: Based on the values of α and ω_0, is the series R-L-C circuit in Figure 17-1 overdamped or underdamped? Does the curve plot confirm this conclusion?

Step 4. Change the value of resistor R in Figure 17-1 to 400 Ω. Click the On-Off switch to run the simulation again. Draw the curve plot of the resistor voltage (circuit current) in the space provided. *Note on the curve plot the time it takes for the current to damp out to zero.*

Step 5. Based on the new value of resistor R in Figure 17-1, calculate the new damping factor (α).

Question: Comparing the new damping factor (α), calculated in Step 5, with the resonant frequency (ω_0), calculated in Step 3, is the R-L-C circuit overdamped or underdamped? Does the curve plot confirm this conclusion?

Step 6. Based on the new damping factor (α), calculated in Step 5, calculate the approximate time it should take for the current to damp out to zero.

Question: Does the curve plot confirm the answer in Step 6?

Step 7. Change the value of resistor R in Figure 17-1 to 200 Ω. Change the Channel B scale to 1 V/Div on the oscilloscope. Click the On-Off switch to run the simulation again. Draw the curve plot of the resistor voltage (circuit current) in the space provided. *Note on the curve plot the time it takes for the current to damp out to zero.* Determine the frequency of oscillation (f_d) from the curve plot and convert it to ω_d. Record the values of f_d and ω_d.

f_d = ______________ ω_d = ______________

Step 8. Based on the new value of resistor R in Figure 17-1, calculate the new damping factor (α).

Step 9. Based on the new damping factor (α), calculated in Step 8, calculate the approximate time it should take for the current to damp out to zero.

Questions: Does the curve plot confirm the answer in Step 9?

What effect does decreasing the value of R have on the damping factor?

What effect does decreasing the value of R have on the R-L-C circuit current curve plot? **Explain.**

Step 10. Based on the damping factor (α), calculated in Step 8, and the resonant frequency (ω_0), calculated in Step 3, calculate the frequency of oscillation (ω_d), in radians per second, for the underdamped R-L-C current curve plot.

Question: How did the calculated value of ω_d in Step 10 compare with the value determined from the curve plot in Step 7?

Step 11. Change the value of capacitor C in Figure 17-1 to 0.05 μF. Click the On-Off switch to run the simulation again. Draw the curve plot of the resistor voltage (circuit current) in the space provided. Determine the frequency of oscillation (f_d) from the curve plot and convert f_d to ω_d. Record the values of f_d and ω_d.

f_d = ______________ ω_d = ______________

Step 12. Based on the new value of capacitor C, calculate the new resonant frequency (ω_0) of the R-L-C circuit, in radians per second.

Step 13. Based on the new damping factor (α) calculated in Step 8 and the new resonant frequency (ω_0) calculated in Step 12, calculate the new frequency of oscillation (ω_d), in radians per second, for the underdamped R-L-C current curve plot.

Questions: Based on the value of f_d in Step 7 and the value of f_d in Step 11, what effect did decreasing the value of C have on the frequency of oscillation for the underdamped R-L-C circuit current?

How did the calculated value of ω_d in Step 13 compare with the value determined from the curve plot in Step 11?

Troubleshooting Problems

1. Pull down the File menu and open FIG17-2. Click the On-Off switch to run the simulation. Based on the current curve plot (blue) on the oscilloscope screen, which circuit component is shorted—L, C, or R? **Explain your choice.**

2. Pull down the File menu and open FIG17-3. Click the On-Off switch to run the simulation. Based on the current curve plot (blue) on the oscilloscope screen, which circuit component is shorted—L, C, or R? **Explain your choice.**

3. Pull down the File menu and open FIG17-4. Click the On-Off switch to run the simulation. Based on the frequency of the damped oscillation, determine the value of C.

   Hint: Measure the time period (T_d) for one cycle of the damped oscillation and calculate the frequency in hertz (f_d) from T_d. Based on the frequency in hertz, calculate the frequency in radians per second (ω_d). Based on the value of ω_d, calculate ω_0.

   C = __________

Part

Alternating Current (AC) Circuits

The experiments in Part IV involve the analysis of ac circuits. You will study ac rms current and voltage, capacitive and inductive reactance, impedance and admittance, ac power and power factor, series and parallel ac circuits, ac nodal and mesh analysis, ac Thevenin equivalent circuits, series and parallel resonance, and passive filters.

The circuits for the experiments in Part IV can be found on the enclosed disk in the PART4 subdirectory.

EXPERIMENT

Name______________________________

Date______________________________

Voltage and Current in AC Circuits—RMS

Objectives:

1. Determine the peak voltage and peak-to-peak voltage of an ac sinusoidal waveshape.
2. Measure the time period for one cycle of an ac sinusoidal voltage and calculate the frequency from the time period.
3. Calculate the angular frequency in radians per second from the frequency in Hz for an ac sinusoidal voltage.
4. Calculate the instantaneous voltage at a particular instant of time from the sinusoidal voltage equation and compare your calculated value with the measured value.
5. Determine the half-cycle and full-cycle average values of an ac sinusoidal voltage.
6. Calculate the rms value of an ac sinusoidal voltage and compare your calculated value with the measured value.
7. Determine the relationship between the ac voltage and the ac current in a resistance element.

Materials:

One dual-trace oscilloscope
One function generator
One 0–10 mA ac milliammeter
One 0–10 V ac voltmeter
Resistors—1 Ω, 1 kΩ

Theory:

With dc voltages and currents, current directions and voltage polarities do not change as a function of time. With ac voltages and currents, current directions and voltage polarities change periodically as a function of time. Although the term **ac** (alternating current) describes any current that periodically changes direction, the term ac is also used to describe any voltage that periodically changes polarity. While ac currents and voltages can follow any periodic waveform pattern, such as a square wave, the term ac most often refers to circuits that involve **sinusoidal** voltages and currents. Sinusoidal voltages and currents follow the waveshape of a sine wave.

In the remaining experiments, you will study circuits that that are energized with sinusoidal voltages and currents. A sinusoidal function is a **periodic** function. The complete transition of the sine wave through one positive alternation and one negative alternation is one **cycle**. The **frequency** (f) of a sinusoidal function, in hertz (Hz), is the number of cycles that a sine wave completes per second. The frequency is the inverse of the time **period** for one cycle (T), in seconds. Therefore,

$$f = \frac{1}{T}$$

The **angular frequency** (ω), in radians per second, can be calculated from

$$\omega = 2\pi f$$

where 2π is the number of radians in one cycle.

The **instantaneous value** of an ac sinusoidal voltage or current can be calculated for a particular value of time (t) from

$$v(t) = V_P \sin \omega t$$

or $$i(t) = I_P \sin \omega t$$

where ωt is the angle in radians, V_P is the peak voltage, and I_P is the peak current. It is assumed that V_P (or I_P) and ω are known. To obtain the angle in degrees from the angle in radians, ωt must be multiplied by $180/\pi$ degrees per radian. It takes 360 degrees, or 2π radians, to complete one sine wave cycle.

The **peak-to-peak** voltage (V_{PP}) or current (I_{PP}) for an ac sinusoidal waveshape is the difference between the positive and negative **peak** values. The peak-to-peak voltage (V_{PP}) or current (I_{PP}) can be calculated from the peak voltage (V_P) or current (I_P) as follows:

$$V_{PP} = 2\ V_P$$

or $$I_{PP} = 2\ I_P$$

where the peak voltage (V_P), or peak current (I_P), is the maximum positive or negative value with respect to zero.

The **average value** of an ac sinusoidal function when measured over a full cycle is zero because the average value of the positive half-cycle is equal to the average value of the negative half-cycle. In order for the average value of an ac sinusoidal function to be useful for comparison purposes, the average value is often defined for one half-cycle. The average value of an ac sinusoidal voltage or current for one half-cycle is calculated from the peak voltage (V_P) or peak current (I_P) as follows:

$$V_{avg} = \frac{2}{\pi} V_P = 0.637 V_P$$

or

$$I_{avg} = \frac{2}{\pi} I_P = 0.637 I_P$$

Because ac voltages and currents vary with time, the power dissipated or generated in an ac circuit also varies with time. Therefore, we are interested in the average power dissipated or generated in an ac circuit. In a resistance element, the power dissipated is proportional to the square of the voltage ($p = v^2/R$) or the square of the current ($p = i^2R$). In an ac circuit, the power dissipated in a resistance element is proportional to the average value of the square (mean square) of the voltage or current. If you take the square root of the mean square of the voltage or current, you get the **root mean square (rms)** value. Therefore, the ac **rms** voltage or current is the ac voltage or current that would dissipate or generate the same average power as the equivalent dc voltage or current. The rms voltage or current is often referred to as the **effective value**. AC voltmeters and ammeters are designed to read rms voltage and current. The ac rms voltage or current for a sinusoidal function can be calculated from the peak voltage or current as follows:

$$V_{rms} = \frac{V_P}{\sqrt{2}} = 0.707V_P$$

$$I_{rms} = \frac{I_P}{\sqrt{2}} = 0.707I_P$$

The circuit in Figure 18-1 will be used to plot the ac sinusoidal voltage across resistor R. From the sinusoidal curve plot, you will measure and calculate the sinusoidal parameters previously discussed. The circuit in Figure 18-2 will be used to demonstrate the relationship between ac voltage and current in a resistance element. The ac voltage across a resistance element is in-phase with the ac resistor current. Ohm's law determines the relationship between the ac rms voltage across the resistor and the ac rms current through the resistor. Therefore,

$$V_{rms} = I_{rms}R$$

Figure 18-1 Sinusoidal Voltage

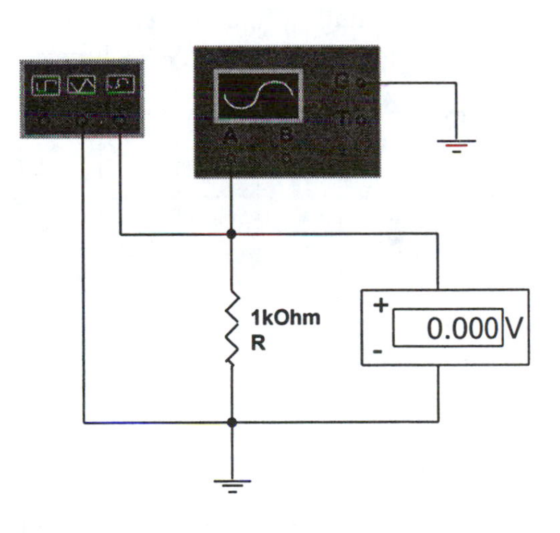

Figure 18-2 Relationship between AC Voltage and Current in a Resistor

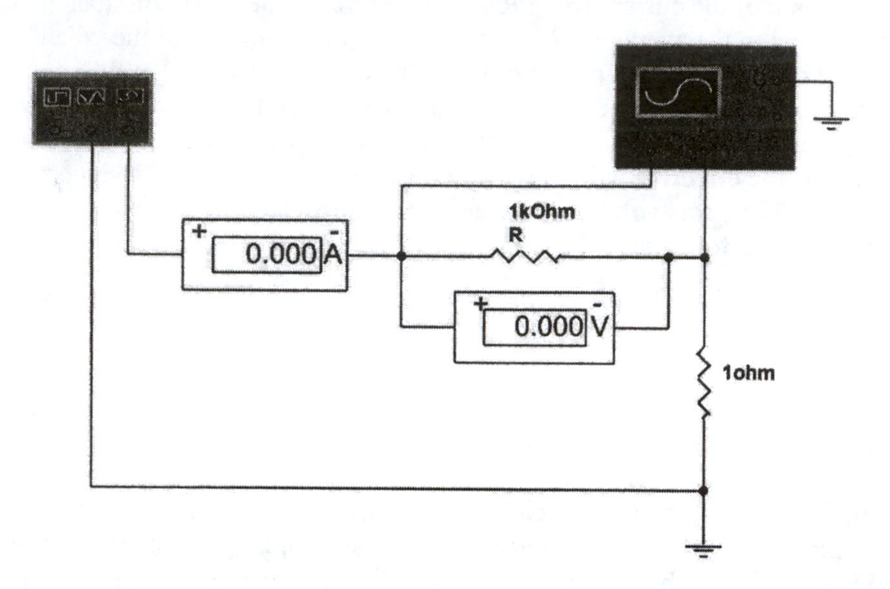

Procedure:

Step 1. Pull down the File menu and open FIG18-1. Bring down the oscilloscope enlargement and make sure that the following settings are selected: Time base (Scale = 1ms/Div, Xpos = 0, Y/T), Ch A (Scale = 5 V/Div, Ypos = 0, AC), Trigger (Pos edge, Level = 1 μV, Sing, A). Bring down the function generator enlargement and make sure that the following settings are selected: *Sine Wave*, Freq = 200 Hz, Ampl = 10 V, Offset = 0. Click the On-Off switch to run the simulation. Draw the ac sinusoidal voltage waveshape in the space provided and *note the peak voltage (V_P) and the time period for one cycle (T) on the curve plot*. Record the ac voltage measured by the AC voltmeter.

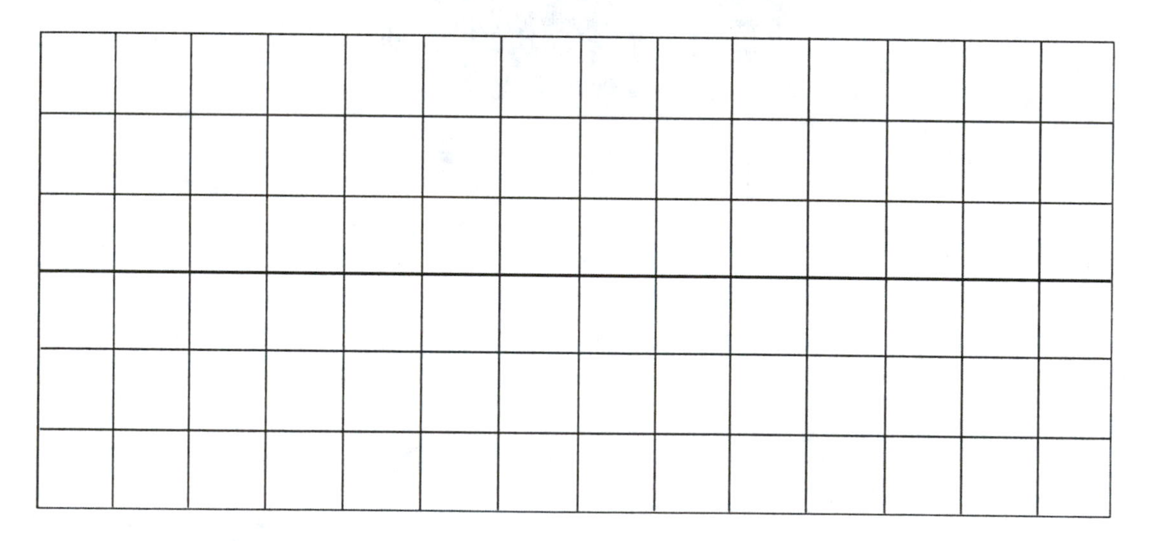

AC voltmeter reading = __________ volts

Step 2. Calculate the peak-to-peak ac voltage (V_{PP}) from the peak voltage (V_P) measured on the curve plot in Step 1.

Step 3. Based on the time period for one cycle (T), calculate the ac sinusoidal frequency (f).

Question: How does your calculated frequency compare with the frequency setting on the function generator?

Step 4. Based on the frequency calculated in Step 3, calculate the angular frequency (ω) in radians per second for the ac sinusoidal voltage.

Step 5. Based on the peak voltage (V_P) on the curve plot in Step 1 and the angular frequency calculated in Step 4, write the equation for the sinusoidal voltage, v(t).

Step 6. Using the equation in Step 5, calculate the instantaneous voltage (v) at t = 2 ms.

Question: How does your calculated instantaneous voltage (v) compare with the instantaneous voltage reading from the curve plot in Step 1 at t = 2 ms?

Step 7. Based on the curve plot in Step 1, calculate the average value of the ac sinusoidal voltage (V_{avg}) for one half-cycle.

Question: What is the average value of the ac sinusoidal voltage for one full cycle? **Explain.**

Step 8. Based on the curve plot in Step 1, calculate the ac rms voltage (V_{rms}).

Question: How did your calculated ac rms voltage compare with the ac voltage reading on the ac voltmeter in Step 1? Was this what you expected? **Explain.**

Step 9. Pull down the File menu and open FIG18-2. Bring down the oscilloscope enlargement and make sure that the following settings are selected: Time base (Scale = 1 ms/Div, Xpos = 0, Y/T), Ch A (Scale = 5 V/Div, Ypos = 0, AC), Ch B (Scale = 10 mV/Div, Ypos = 0, AC), Trigger (Pos edge, Level = 1 μV, Sing, A). Bring down the function generator enlargement and make sure that the following settings are selected: *Sine Wave*, Freq = 200 Hz, Ampl = 10 V, Offset = 0. Click the On-Off switch to run the simulation. The voltage across the 1 Ω resistor is proportional to the current in the 1 kΩ resistor (blue curve plot). Because the 1 kΩ resistor is so much higher in value than the 1 Ω resistor, the voltage across both resistors (red curve plot) is practically equal to the voltage across the 1 kΩ resistor. Therefore, the red curve is plotting the voltage across the 1 kΩ resistor and the blue curve is plotting the current in the 1 kΩ resistor (1 mV is equivalent to 1 mA). Draw the curve plots for the resistor voltage and current in the space provided on the following page and **note the peak values on the diagram**. Record the ac voltmeter and ac ammeter readings.

AC voltmeter reading = __________ AC ammeter reading = __________

Question: What is the phase relationship between the resistor voltage and current?

Step 10. Based on the peak voltage (V_P) of the red curve plot in Step 9, calculate the rms voltage (V_{rms}) across the resistor.

Question: How did your calculated ac rms voltage compare with the voltage measured on the ac voltmeter in Step 9?

Step 11. Based on the peak voltage across the 1 Ω resistor (blue curve plot) in Step 9, calculate the peak current (I_P) and the rms current (I_{rms}).

Question: How did your calculated ac rms current compare with the current measured on the ac ammeter in Step 9?

Step 12. Based on the rms current and voltage calculated in Steps 10 and 11, calculate the resistance of resistor R.

Questions: How did your calculated value for resistor R compare with the actual value of R? Did this prove that Ohm's law applies to ac resistance circuits?

Troubleshooting Problem

1. Pull down the File menu and open FIG18-3. Click the On-Off switch to run the simulation. Based on the resistor voltage and current curve plots on the oscilloscope, determine the value of R.

   R = __________

EXPERIMENT

19

Name________________________________

Date________________________________

Inductive Reactance

Objectives:

1. Determine the phase relationship between the ac voltage and current in an inductor.
2. Determine the inductive reactance of an inductor from the measured ac rms inductor voltage and current and compare this value to the calculated value.
3. Determine the relationship between the inductive reactance of an inductor and the value of the inductance.
4. Determine the relationship between the inductive reactance of an inductor and the ac sinusoidal frequency.

Materials:

One dual-trace oscilloscope
One function generator
One 0–15 mA ac milliammeter
One 0–10 V ac voltmeter
One 1 Ω resistor
Inductors—100 mH, 50 mH

Theory:

An ac sinusoidal current has its maximum rate of change at its zero crossing points and its minimum rate of change at its peaks. From **Faraday's law**, the voltage induced across the terminals of an inductor is proportional to the rate of change of the inductor current. When an ac sinusoidal current is passed through an inductor, the **induced voltage** is maximum when the inductor current is at a zero crossing point and minimum when the inductor current is at a peak. The induced voltage is positive when the current is rising and negative when the current is falling. Therefore, in a pure inductor (neglecting any inductor resistance, R_L) the ac inductor current (I_L) **lags** the ac inductor voltage (V_L) by **90 degrees** and the phase difference is not dependent on the inductor value (L) or the ac frequency (f). This will be demonstrated with the circuit in Figure 19-1.

From **Lenz's law**, a changing inductor current induces a voltage across an inductor in a direction that opposes the change in the current. When an ac sinusoidal current is applied to an inductor, a voltage is induced in a direction to always oppose the change in the ac current. This opposition to the change in the ac current in an inductor is similar to the opposition to dc current in a resistance element. This resistance to the change in the ac inductor current is called **inductive reactance (X_L),** and is measured in ohms (Ω). As in a dc resistive circuit, **Ohm's law** determines the relationship between the ac rms inductor voltage (V_L) and the ac rms inductor current (I_L). In an ac circuit, the inductive reactance (X_L) of an inductor replaces the resistance of a resistor in the Ohm's law equation. Therefore,

$$X_L = \frac{V_L}{I_L}$$

Because the rms voltage and current are both equal to the peak value times 0.707, Ohm's law can also be applied to the peak inductor voltage (V_P) and peak inductor current (I_P) as follows:

$$X_L = \frac{0.707V_P}{0.707I_P} = \frac{V_P}{I_P}$$

Because the induced voltage across an inductor is proportional to the rate of change of the current (Faraday's law) and the inductance (L), the inductive reactance (X_L) of an inductor is proportional to the ac sinusoidal frequency (f) and the inductance (L). Therefore, the inductive reactance (X_L) can be calculated from

$$X_L = \omega L = 2\pi fL$$

where ω is the frequency in radians per second, f is the frequency in hertz (Hz), L is the inductance in henries (H), and X_L is the inductive reactance in ohms (Ω).

If you perform this experiment in a real lab environment, you will obtain ac current readings slightly less than the readings obtained in the computer simulation because a real inductor has coil wire resistance. If the inductive reactance (X_L) is much greater than the resistance of the inductor wire (R_L), the results will be very close to the computer simulated results. If closer results are desired, add a resistor of value R_L in series with the inductor in the computer simulated circuit.

Figure 19-1 Inductive Reactance

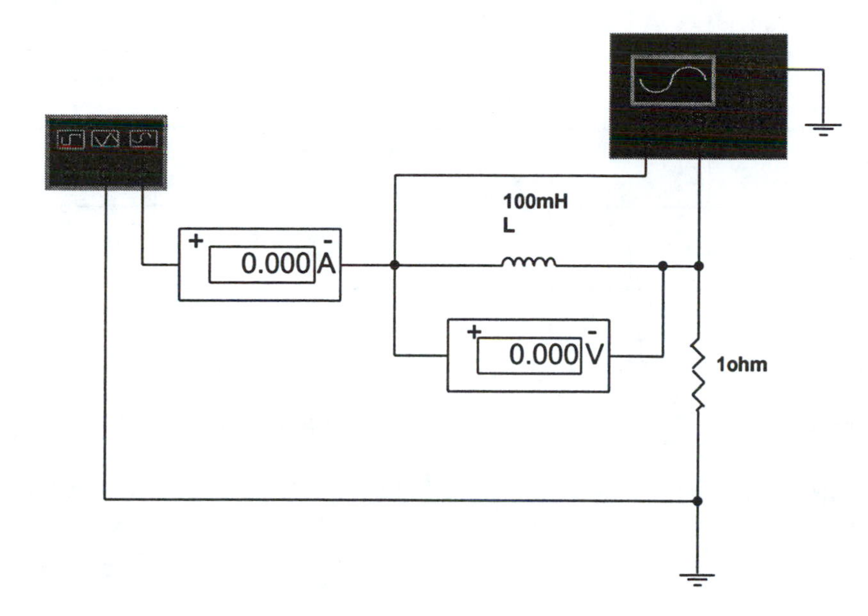

Procedure:

Step 1. Pull down the File menu and open FIG19-1. Bring down the oscilloscope enlargement and make sure that the following settings are selected: Time base (Scale = 50 μs/Div, Xpos = 0, Y/T), Ch A (Scale = 5 V/Div, Ypos = 0, AC), Ch B (Scale = 2 mV/Div, Ypos = 0, AC), Trigger (Pos edge, Level =1 μV, Nor, A). Bring down the function generator enlargement and make sure that the following settings are selected: *Sine Wave*, Freq = 5 kHz, Ampl = 10 V, Offset = 0. Click the On-Off switch to run the simulation. Click "pause" after three screen displays on the oscilloscope. The ac voltage across the 1 Ω resistor is proportional to the ac current in the inductor (blue curve plot). Because the inductive reactance is so much higher in value than the 1 Ω resistor, the ac voltage across the inductor and the 1 Ω resistor (red curve plot) is practically equal to the ac voltage across the inductor. Therefore, the red curve is plotting the ac voltage across the inductor (V_L) and the blue curve is plotting the ac current in the inductor (I_L) (1 mV is equivalent to 1 mA). Draw the curve plots for the inductor ac voltage (V_L) and ac current (I_L) in the space provided on the following page and *note the peak values on the diagram.* Record the ac voltmeter and ac ammeter readings.

AC voltmeter reading = __________ AC ammeter reading = __________

Question: What is the phase relationship between the inductor voltage and current? Which leads and which lags in phase? By how many degrees?

Step 2. Based on the peak inductor voltage (V_P) on the curve plot in Step 1, calculate the rms voltage across the inductor (V_L).

Question: How did your calculated ac rms inductor voltage (V_L) compare with the inductor voltage measured on the ac voltmeter in Step 1? **Explain.**

Step 3. Based on the peak inductor current (I_P) on the curve plot in Step 1, calculate the rms inductor current (I_L).

Question: How did your calculated ac rms inductor current (I_L) compare with the current measured on the ac ammeter in Step 1?

Step 4. Based on the ac rms inductor voltage and current readings in Step 1, determine the inductive reactance (X_L) of inductor L.

Step 5. Based on the ac sinusoidal frequency (f) and the value of the inductance (L), calculate the expected inductive reactance (X_L).

Question: How did your calculated inductive reactance (X_L) in Step 5 compare with the inductive reactance determined from the measured inductor ac rms voltage and current in Step 4?

Step 6. Change the inductance (L) to 50 mH. Change the Channel B Scale to 5 mV/Div on the oscilloscope. Click the On-Off switch to run the simulation. Click "Pause" after three screen displays on the oscilloscope. Record the ac rms voltmeter and ammeter readings.

AC voltmeter reading = __________ AC ammeter reading = __________

Question: Did the phase relationship between the inductor voltage and current curve plots on the oscilloscope change? **Explain.**

Step 7. Based on the new ac rms inductor voltage and current readings in Step 6, determine the new inductive reactance (X_L).

Question: How did the new value for the inductive reactance (X_L) compare with the inductive reactance when the inductance (L) was 100 mH in Step 4? **Explain.**

Step 8. Based on the ac sinusoidal frequency (f) and the new value of the inductance (L), calculate the expected new inductive reactance (X_L).

Question: How did your calculated inductive reactance (X_L) in Step 8 compare with the inductive reactance determined from the measured inductor ac rms voltage and current in Step 7?

Step 9. Change the inductance (L) back to 100 mH. Change the frequency (f) of the function generator to 10 kHz. Change the Channel B input on the oscilloscope to 1 mV/Div. Click the On-Off switch to run the simulation. Click "Pause" after three screen displays on the oscilloscope. Record the ac rms voltmeter and ammeter readings.

AC voltmeter reading = __________ AC ammeter reading = __________

Question: Did the phase relationship between the inductor voltage and current curve plots on the oscilloscope change? **Explain.**

Step 10. Based on the new ac rms inductor voltage and current readings in Step 9, determine the new inductive reactance (X_L).

Question: How did the new value for the inductive reactance (X_L) compare with the inductive reactance when the frequency (f) was 5 kHz in Step 4? **Explain.**

Step 11. Based on the new ac sinusoidal frequency (f) and the value of the inductance (L), calculate the expected new inductive reactance (X_L).

Question: How did your calculated inductive reactance (X_L) in Step 11 compare with the inductive reactance determined from the measured inductor ac rms voltage and current in Step 10?

Step 12. In Table 19-1, record the values for the rms inductor voltage (V_L), rms inductor current (I_L), and inductive reactance (X_L) already measured and calculated at 5 kHz and 10 kHz in Steps 1, 4, 9, and 10. Then set the function generator to each other frequency in Table 19-1, run the simulation, and record the ac rms inductor voltage and current for each frequency.

Table 19-1

f (kHz)	V_L (V)	I_L (mA)	X_L (kΩ)
1			
2			
5			
10			

Step 13. Calculate the inductive reactance (X_L) for each frequency from the ac rms inductor voltage (V_L) and current (I_L) and record your answers in Table 19-1.

Step 14. Based on the data in Table 19-1, plot a graph of the inductive reactance (X_L) versus frequency (f).

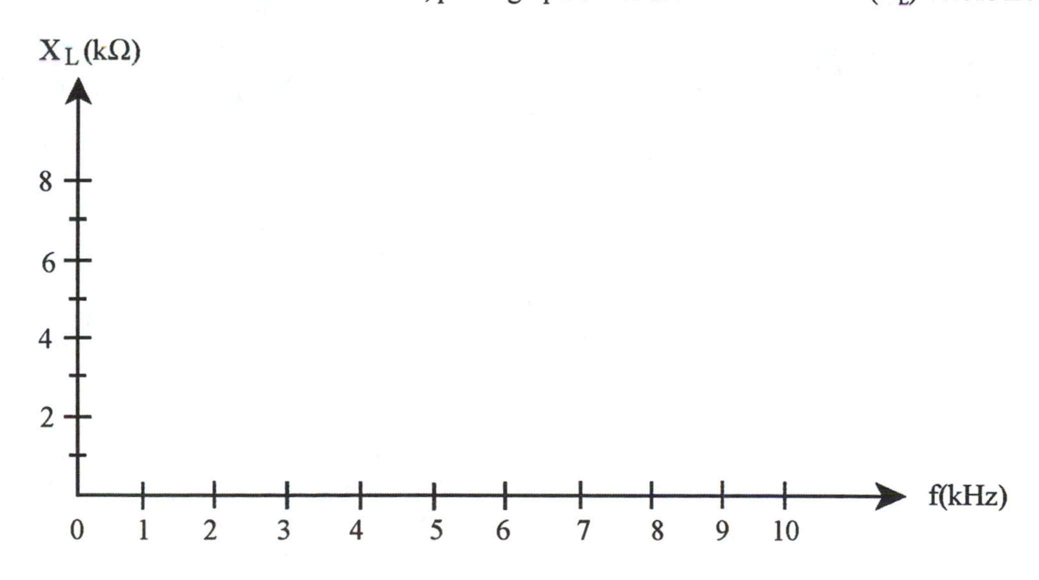

Question: Did the inductive reactance (X_L) increase or decrease when the frequency was increased? Was that what you expected? **Explain.**

Troubleshooting Problem

1. Pull down the File menu and open FIG19-2. Click the On-Off switch to run the simulation. Based on the ac meter readings, determine the value of L.

   L = __________

Name______________________________

Date______________________________

Capacitive Reactance

Objectives:

1. Determine the phase relationship between the ac voltage and current in a capacitor.
2. Determine the capacitive reactance of a capacitor from the measured ac rms capacitor voltage and current and compare this value to the calculated value.
3. Determine the relationship between the capacitive reactance of a capacitor and the value of the capacitance.
4. Determine the relationship between the capacitive reactance of a capacitor and the ac sinusoidal frequency.

Materials:

One dual-trace oscilloscope
One function generator
One 0–15 mA ac milliammeter
One 0–10 V ac voltmeter
One 1 Ω resistor
Capacitors—0.05 μF, 0.1 μF

Theory:

An ac sinusoidal voltage has its maximum rate of change at its zero crossing points and its minimum rate of change at its peaks. The **charging current** in a capacitor is proportional to the rate of change of the capacitor voltage. When an ac sinusoidal voltage is applied across a capacitor, the charging current is maximum when the capacitor voltage is at a zero crossing point and minimum when the capacitor voltage is at a peak. The charging current is positive when the voltage is rising and negative when the voltage is falling. Therefore, in a capacitor the ac capacitor voltage (V_C) **lags** the ac capacitor current (I_C) by **90 degrees** and the phase difference is not dependent on the capacitance (C) or the ac frequency (f). This will be demonstrated with the circuit in Figure 20-1.

When an ac sinusoidal current is applied to a capacitor, the capacitor voltage is in a direction to always oppose the change in the ac current. This opposition to the change in the ac current in a capacitor is similar to the opposition to dc current in a resistance element. This resistance to the change in the ac capacitor current is called **capacitive reactance (X_C),** and is measured in ohms **(Ω)**. As in a dc resistive circuit, **Ohm's law** determines the relationship between the ac rms capacitor voltage (V_C) and the ac rms capacitor current (I_C). In an ac circuit, the capacitive reactance (X_C) of a capacitor replaces the resistance of a resistor in the Ohm's law equation. Therefore,

$$X_C = \frac{V_C}{I_C}$$

Because the rms voltage and current are both equal to the peak value times 0.707, Ohm's law can also be applied to the peak capacitor voltage (V_P) and peak capacitor current (I_P) as follows:

$$X_C = \frac{0.707V_P}{0.707I_P} = \frac{V_P}{I_P}$$

Because the capacitor current decreases when the capacitor voltage increases, and is proportional to the capacitance (C) and the rate of change of the capacitor voltage, the capacitive reactance (X_C) of a capacitor is inversely proportional to the ac sinusoidal frequency (f) and the capacitance (C). Therefore, the capacitive reactance (X_C) can be calculated from

$$X_C = \frac{1}{\omega C} = \frac{1}{2\pi fC}$$

where ω is the frequency in radians per second, f is the frequency in hertz (Hz), C is the capacitance in farads (F), and X_C is the capacitive reactance in ohms (Ω).

Figure 20-1 Capacitive Reactance

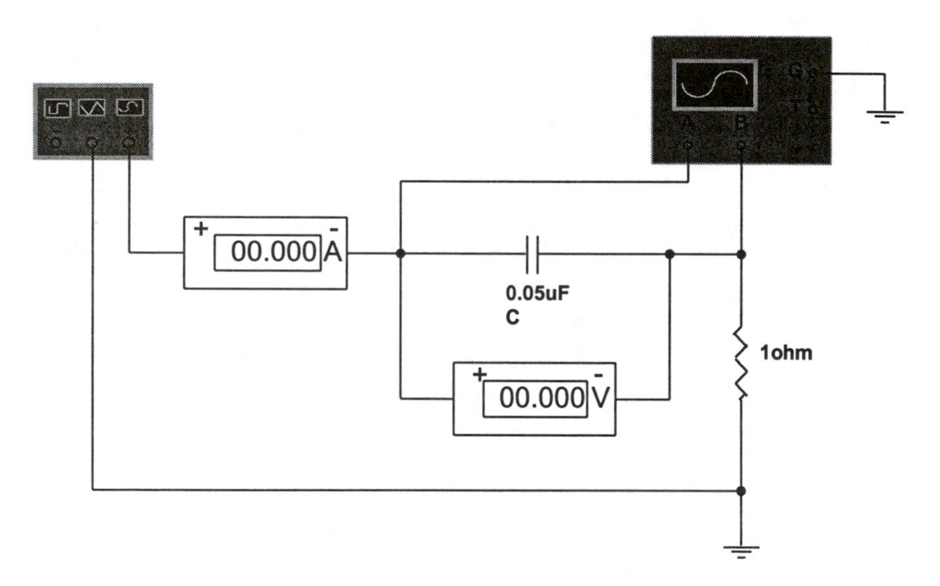

Procedure:

Step 1. Pull down the File menu and open FIG20-1. Bring down the oscilloscope enlargement and make sure that the following settings are selected: Time base (Scale = 200 μs/Div, Xpos = 0, Y/T), Ch A (Scale = 5/Div, Ypos = 0, AC), Ch B (Scale = 2 mV/Div, Ypos = 0, AC), Trigger (Pos edge, Level = 1 μV, Sing A). Bring down the function generator enlargement and make sure that the following settings are selected: *Sine Wave*, Freq = 1 kHz, Ampl = 10 V, Offsett = 0. Click the On-Off switch to run the simulation. The ac voltage across the 1 Ω resistor is proportional to the ac current in the capacitor (blue curve plot). Because the capacitive reactance is so much higher in value than the 1 Ω resistor, the ac voltage across the capacitor and the 1 Ω resistor (red curve plot) is practically equal to the ac voltage across the capacitor. Therefore, the red curve is plotting the ac voltage across the capacitor (V_C) and the blue curve is plotting the ac current in the capacitor (I_C) (1 mV is equivalent to 1 mA). Draw the curve plots for the capacitor ac voltage (V_C) and ac current (I_C) in the space provided and *note the peak values on the diagram*. Record the ac voltmeter and ac ammeter readings.

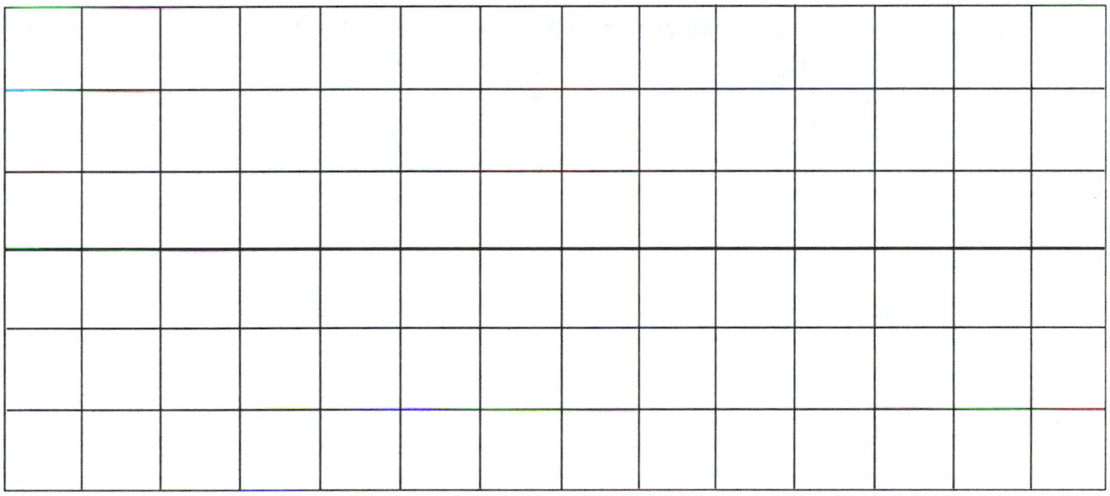

AC voltmeter reading = __________ AC ammeter reading = __________

Question: What is the phase relationship between the capacitor voltage and current? Which leads and which lags in phase? By how many degrees?

Step 2. Based on the peak capacitor voltage (V_P) on the curve plot in Step 1, calculate the rms voltage across the capacitor (V_C).

Question: How did your calculated ac rms capacitor voltage (V_C) compare with the capacitor voltage measured on the ac voltmeter in Step 1? **Explain.**

Step 3. Based on the peak capacitor current (I_P) on the curve plot in Step 1, determine the rms capacitor current (I_C).

Question: How did your calculated ac rms capacitor current (I_C) compare with the current measured on the ac ammeter in Step 1?

Step 4. Based on the ac rms capacitor voltage and current readings in Step 1, determine the capacitive reactance (X_C) of capacitor C.

Step 5. Based on the ac sinusoidal frequency (f) and the value of the capacitance (C), calculate the expected capacitive reactance (X_C).

Question: How did your calculated capacitive reactance (X_C) in Step 5 compare with the capacitive reactance determined from the measured capacitor ac rms voltage and current in Step 4?

Step 6. Change the capacitance (C) to 0.1 μF. Change the Channel B Scale to 5 mV/Div on the oscilloscope. Click the On-Off switch to run the simulaiton. Record the ac voltmeter and ammeter readings.

AC voltmeter reading = __________ AC ammeter reading = __________

Question: Did the phase relationship between the capacitor voltage and current curve plots on the oscilloscope change? **Explain.**

Step 7. Based on the new ac rms capacitor voltage and current readings in Step 6, determine the new capacitive reactance (X_C).

Question: How did the new value for the capacitive reactance (X_C) compare with the capacitive reactance when the capacitance (C) was 0.05 μF in Step 4? **Explain.**

Step 8. Based on the ac sinusoidal frequency (f) and the new value of the capacitance (C), calculate the expected new capacitive reactance (X_C).

Question: How did your calculated capacitive reactance (X_C) in Step 8 compare with the capacitive reactance determined from the measured capacitor ac rms voltage and current in Step 7?

Step 9. Change the capacitance (C) back to 0.05 μF. Change the frequency (f) of the function generator to 2 kHz. Click the On-Off switch to run the simulation. Record the ac voltmeter and ammeter readings.

AC voltmeter reading = __________ AC ammeter reading = __________

Question: Did the phase relationship between the capacitor voltage and current curve plots on the oscilloscope change? **Explain.**

Step 10. Based on the new ac rms capacitor voltage and current readings in Step 9, determine the new capacitive reactance (X_C).

Question: How did the new value for the capacitive reactance (X_C) compare with the capacitive reactance when the frequency (f) was 1 kHz in Step 4? **Explain.**

Step 11. Based on the new ac sinusoidal frequency (f) and the value of the capacitance (C), calculate the expected new capacitive reactance (X_C).

Question: How did your calculated capacitive reactance (X_C) in Step 11 compare with the capacitive reactance determined from the measured capacitor ac rms voltage and current in Step 10?

Step 12. In Table 20-1, record the values for the ac rms capacitor voltage (V_C), ac rms capacitor current (I_C), and capacitive reactance (X_C) already measured and calculated at 1 kHz and 2 kHz in Steps 1, 4, 9, and 10. Then set the function generator to each other frequency in Table 20-1, run the simulation, and record the ac rms capacitor voltage and current for each frequency.

Table 20-1

f (kHz)	V_C (V)	I_C (mA)	X_C (kΩ)
1			
2			
3			
4			
5			

13. Calculate the capacitive reactance (X_C) for each frequency from the ac rms capacitor voltage (V_C) and current (I_C) and record your answers in Table 20-1.

14. Based on the data in Table 20-1, plot a graph of the capacitive reactance (X_C) versus frequency (f).

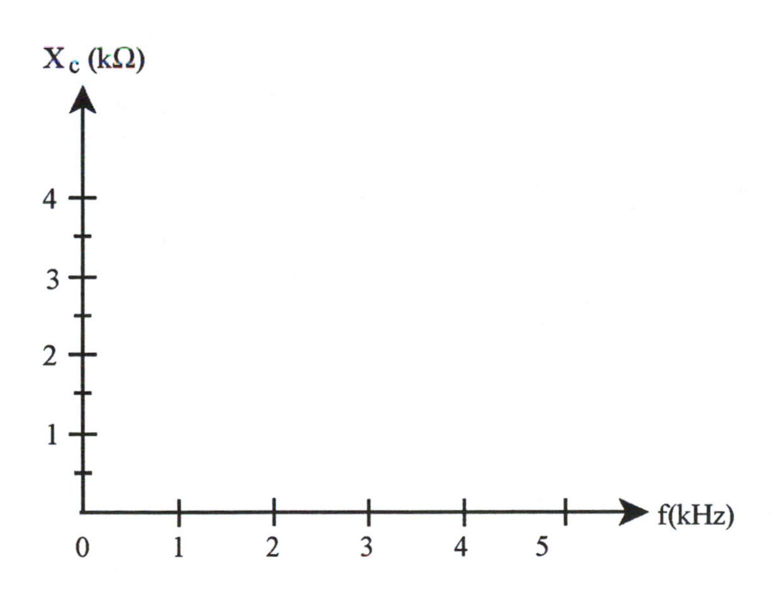

Question: Did the capacitive reactance (X_C) increase or decrease when the frequency was increased? Was that what you expected? **Explain**.

Troubleshooting Problem

1. Pull down the File menu and open FIG20-2. Click the On-Off switch to run the simulation. Based on the ac meter readings, determine the value of C.

   C = __________

EXPERIMENT

Name______________________________

Date______________________________

Impedance of Series AC Circuits

Objectives:

1. Measure the impedance of a series R-L circuit and the phase between the ac voltage and current and compare your measured values with the calculated values.
2. Measure the impedance of a series R-C circuit and the phase between the ac voltage and current and compare your measured values with the calculated values.
3. Measure the impedance of a series R-L-C circuit and the phase between the ac voltage and current and compare your measured values with the calculated values.
4. Calculate the expected ac rms voltage across each element in a series R-L-C circuit and compare your calculated voltages with the measured values.
5. Demonstrate how Kirchhoff's voltage law applies to a series ac impedance using phasors.
6. Demonstrate the effect of frequency changes on the ac rms current and voltages in a series R-L-C circuit.

Materials:

One dual-trace oscilloscope
One function generator
One 0–10 mA ac milliammeter
Four 0–10 V ac voltmeters
One 0.1 μF capacitor
One 100 mH inductor
One 1 kΩ resistor

Theory:

Complete Experiments 19 and 20 on inductive and capacitive reactance before attempting this experiment.

When an ac sinusoidal voltage is applied across a series R-L, R-C, or R-L-C circuit, as shown in Figures 21-1, 21-2, and 21-3, there is an opposition to the ac current flow called **impedance (Z)**, and its unit of measurement is the ohm (Ω). The voltage across each element and the current in each element are also sinusoidal and have the same frequency as the applied voltage. However, the ac voltage across the inductor will lead the ac current by 90 degrees, the ac voltage across the capacitor will lag the ac current by 90 degrees, and the ac voltage across the resistor will be in-phase with the ac current. This will cause a **phase difference (θ)** between the ac voltage applied to the circuit and the ac circuit current. This phase difference can be between **0 degrees and 90 degrees**, depending on the relationship between the total **reactance** and total **resistance** in the circuit.

As in a dc resistive circuit, **Ohm's law** determines the relationship between the ac rms voltage applied to the series circuit and the ac rms circuit current. In an ac circuit, the impedance (Z) of the circuit replaces the resistance of a resistor in the Ohm's law equation. Therefore,

$$V = Z I$$

where Z is the circuit impedance in ohms (Ω), V is the ac rms voltage applied to the circuit, and I is the ac rms circuit current. Because the rms voltage and current are both equal to the peak value times 0.707, Ohm's law can also be applied to the peak ac voltage (V_P) and peak ac current (I_P) as follows:

$$V_P = Z I_P$$

Ohm's law can be used to find the voltage across an inductive reactance (X_L), a capacitive reactance (X_C), and a resistance (R). Therefore,

$$V_L = IX_L$$

for an inductor,

$$V_C = IX_C$$

for a capacitor, and

$$V_R = IR$$

for a resistor.

When applying **Kirchhoff's voltage law** to a series circuit ac impedance, the ac voltages must be added using **phasor addition** because they are out-of-phase with each other. Both the magnitude and the phase of each voltage must be taken into account.

The ac voltage across the **circuit reactance** (IX) is **90 degrees out-of-phase** with the ac circuit current, and the ac voltage across the **circuit resistance** (IR) is **in phase** with the ac circuit current. Therefore, IR and IX must be added as if they are separated by 90 degrees. From right angle trigonometry (Pythagorean theorem), Kirchhoff's voltage law, and Ohm's law, the ac rms voltages around the closed path can be represented by

$$V = \sqrt{(IR)^2 + (IX)^2} = \sqrt{I^2(R^2 + X^2)} = I\sqrt{R^2 + X^2} = IZ$$

Based on this equation, the series circuit impedance (Z) can be determined from

$$Z = \sqrt{R^2 + X^2}$$

where X is the circuit reactance in ohms, and R is the circuit resistance in ohms.

From right angle trigonometry (Pythagorean theorem), the **phase difference** between the ac rms voltage applied across the impedance and the ac rms circuit current can be calculated from

$$\theta = \arctan\left(\frac{IX}{IR}\right) = \arctan\left(\frac{X}{R}\right)$$

where θ is the phase difference in radians or degrees. As can be seen from the equation, the larger the reactance (X) compared to the resistance (R), the larger the phase difference (θ) between the ac rms voltage and current.

The phase difference (θ) between two periodic functions (such as sine functions) can be measured by measuring the time difference (t) between the two curve plots and the time period for one cycle (T) of the curve plots. Because the ratio of the time difference (t) divided by the time period (T) for one cycle is equal to the ratio of the phase difference (θ), in degrees, divided by the number of degrees of phase for one cycle (360 degrees), the phase difference can be calculated from

$$\frac{\theta}{360^\circ} = \frac{t}{T}$$

Therefore,

$$\theta = \left(\frac{t}{T}\right)360^\circ$$

The impedance (Z) of the series R-L ac circuit in Figure 21-1 is the phasor sum of the resistance (R) and the inductive reactance (X_L). Therefore, the magnitude of the impedance (Z) can be found from

$$Z = \sqrt{R^2 + X_L^2}$$

and the phase difference (θ) between the ac rms voltage across the impedance (V) and the ac rms current (I) can be found from

$$\theta = \arctan\left(\frac{X_L}{R}\right)$$

***Note*: See the Theory section in Experiment 19 for calculating inductive reactance (X_L).**

The impedance (Z) of the series R-C ac circuit in Figure 21-2 is the phasor sum of the resistance (R) and the capacitive reactance (X_C). Therefore, the magnitude of the impedance (Z) can be found from

$$Z = \sqrt{R^2 + X_C^2}$$

and the phase difference (θ) between the ac rms voltage across the impedance (V) and the ac rms current (I) can be found from

$$\theta = -\arctan\left(\frac{X_C}{R}\right)$$

Note that when the voltage lags the current, the phase difference (θ) is negative.

***Note*: See the Theory section in Experiment 20 for calculating capacitive reactance (X_C).**

The impedance (Z) of the series R-L-C ac circuit in Figure 21-3 is the phasor sum of the resistance (R) and the **total reactance** (X) of the inductor and capacitor. The total reactance (X) is equal to the phasor sum of the **inductive reactance** (X_L) and the **capacitive reactance** (X_C). Because the inductive reactance and capacitive reactance phasors are 180 degrees out of phase, the total reactance (X) is

$$X = X_L - X_C$$

Therefore, the magnitude of the impedance (Z) for the series R-L-C circuit can be found from

$$Z = \sqrt{R^2 + X^2}$$

and the phase difference (θ) between the ac rms voltage across the impedance (V) and the ac rms current (I) can be found from

$$\theta = \arctan\left(\frac{X}{R}\right)$$

The inductive reactance (X_L) and the capacitive reactance (X_C) are a function of the ac sinusoidal frequency. In a series R-L-C ac circuit, there is only one frequency at which they are equal. At this frequency, the total reactance is zero ($X = X_L - X_C = 0$) and the circuit impedance is resistive and at the minimum value.

In the series R-L-C circuit in Figure 21-3, the ac rms voltage across the resistance (V_R) is in-phase with the ac rms circuit current (I), the ac rms voltage across the inductor (V_L) leads the circuit current (I) by 90 degrees, and the ac rms voltage across the capacitor (V_C) lags the circuit current (I) by 90 degrees. The phasor diagram representing the voltage phasors is shown on the next page. Because phasors V_L and V_C are 180 degrees out-of-phase, the voltage across the total reactance (V_X) is equal to the difference between V_L and V_C. Therefore,

$$V_X = V_L - V_C$$

Notice that the phasor sum of V_R and V_X is equal to V. From the triangle,

$$V = \sqrt{V_R^2 + V_X^2}$$

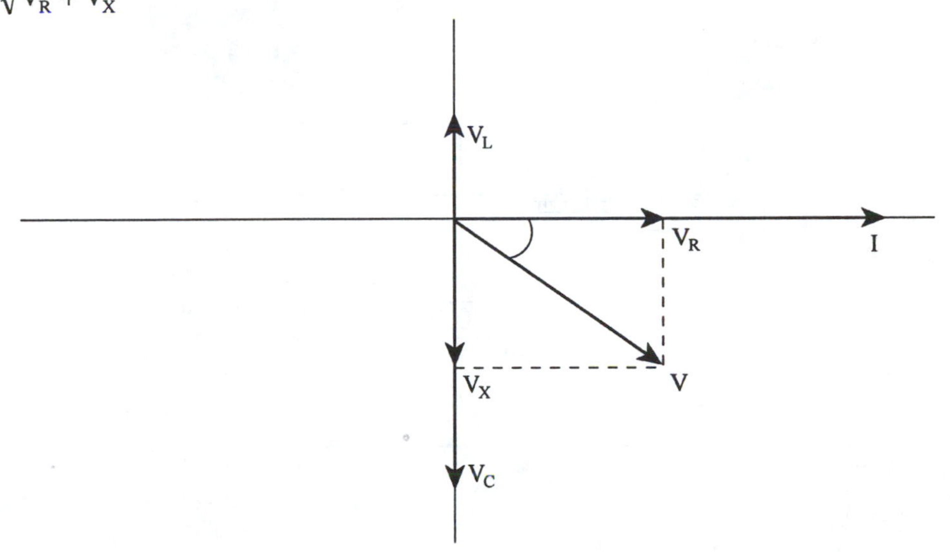

> *Note:* If you do the experiment in a real laboratory environment (hardwired), see the end of the Theory section of Experiment 19 regarding the inductor resistance (R_L).

Figure 21-1 Impedance—Series R-L Circuit

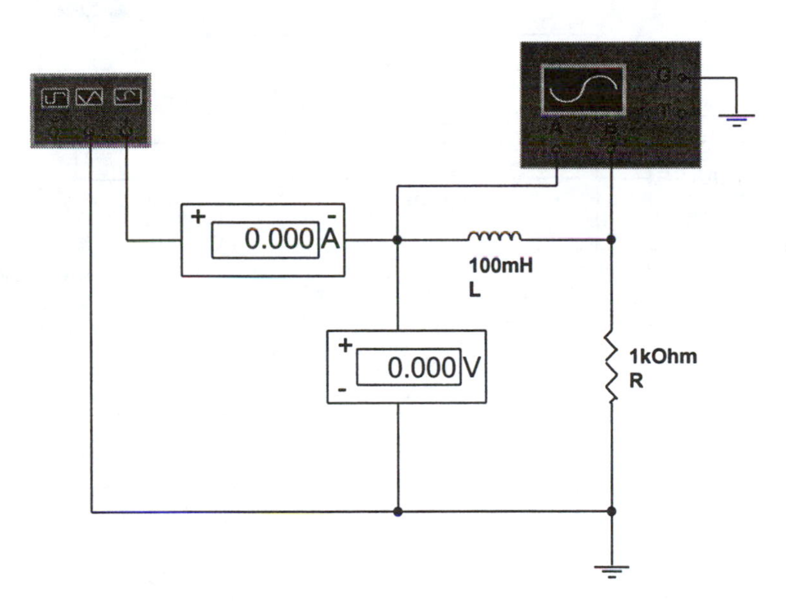

Figure 21-2 Impedance—Series R-C Circuit

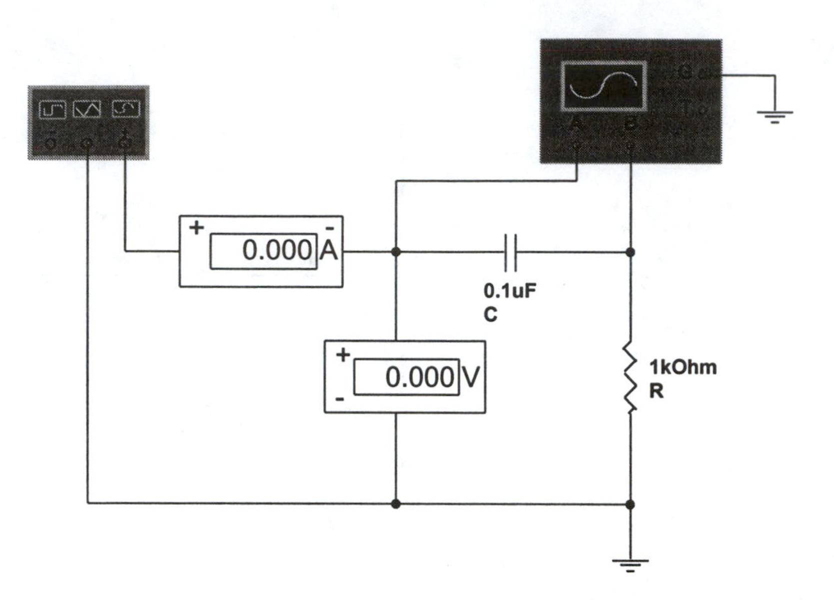

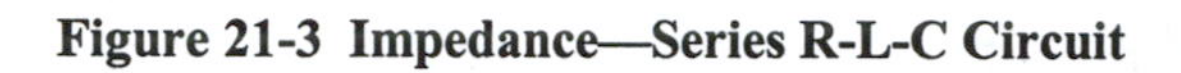

Figure 21-3 Impedance—Series R-L-C Circuit

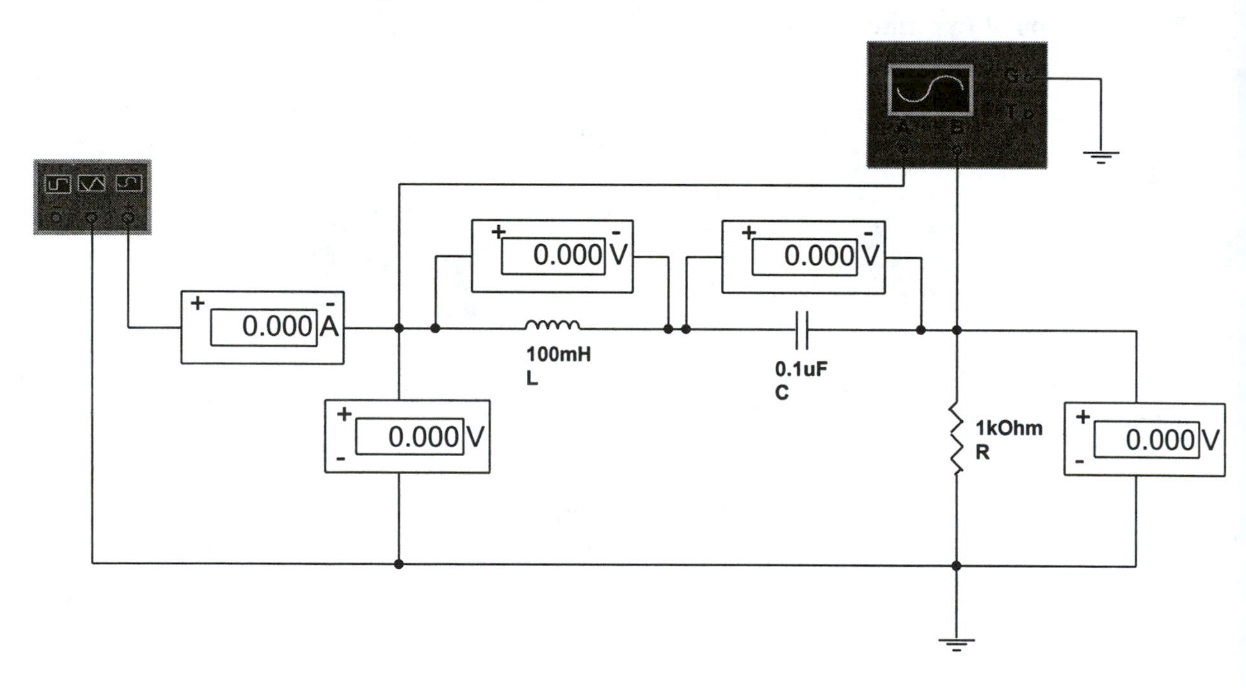

Procedure:

Step 1. Pull down the File menu and open FIG21-1. Bring down the oscilloscope enlargement and make sure that the following settings are selected: Time base (Scale = 100 μs/Div, Xpos = 0, Y/T), Ch A (Scale = 5 V/Div, Ypos = 0, AC), Ch B (Scale = 5 V/Div, Ypos = 0, AC), Trigger (Pos edge, Level = 1 μV, Nor, A). Bring down the function generator enlargement and make sure that the following settings are selected: *Sine Wave*, Freq = 2 kHz, Ampl = 10 V, Offset = 0. Click the On-Off switch to run the simulation. Click "Pause" after ten screen displays on the oscilloscope. The red curve is plotting the voltage across the impedance of the R-L circuit (V), and the blue curve is plotting the current (I) because the voltage across the 1 kΩ resistor is proportional to the current (1 V is equivalent to 1 mA on the oscilloscope vertical axis). Draw the curve plots for the voltage (V) and current (I) in the space provided. Record the ac rms voltage (V) and current (I) readings on the ac voltmeter and ammeter.

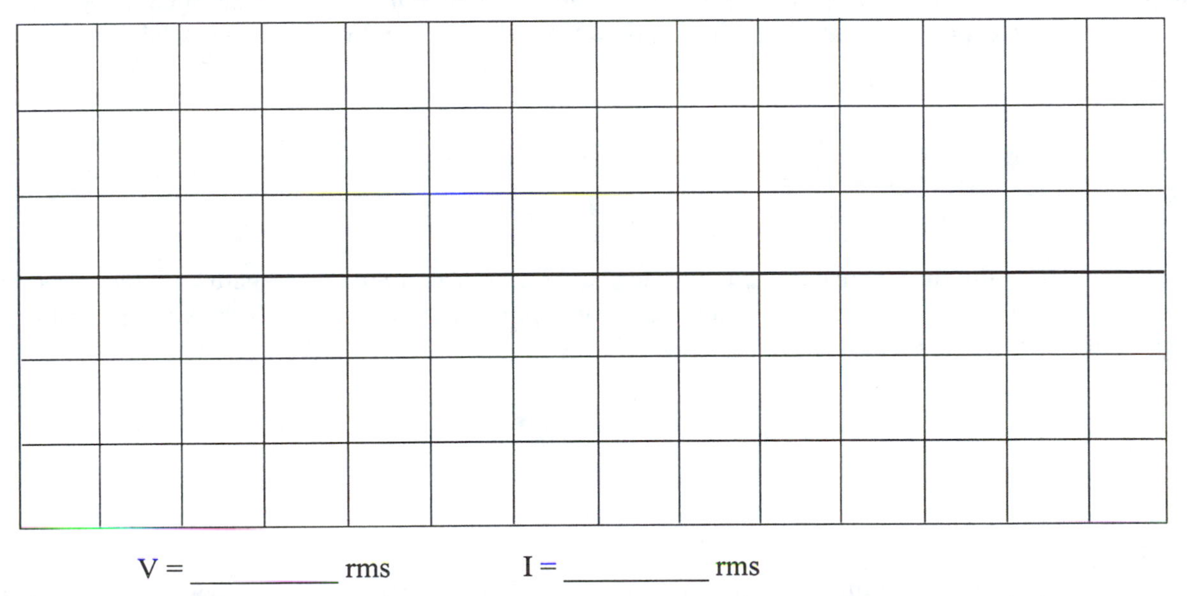

V = __________ rms I = __________ rms

Step 2. Based on the curve plots in Step 1, determine the phase difference (θ) between the voltage and current.

Step 3. Based on the ac rms voltage (V) and current (I), calculate the magnitude of the impedance (Z) of the R-L circuit.

Step 4. Based on the value of the inductance (L) and the sinusoidal frequency (f), calculate the inductive reactance (X_L) of the inductor.

Step 5. Based on the value of resistance R and the inductive reactance (X_L) of inductor L, calculate the expected magnitude of the impedance (Z) of the R-L circuit.

Question: How did your calculated impedance magnitude in Step 5 compare with the impedance from the measured ac rms voltage and current in Step 3?

Step 6. Based on the calculated inductive reactance (X_L) and the value of resistance R, calculate the expected phase difference (θ) between the current and voltage sinusoidal functions.

Question: How did the calculated value for the phase difference in Step 6 compare with the measured phase difference between the current and voltage curve plots in Steps 1 and 2? Is the voltage leading or lagging the current? Is this what you expected?

Step 7. Pull down the File menu and open FIG21-2. Bring down the oscilloscope enlargement and make sure that the following settings are selected: Time base (Scale = 100 μs/Div, Xpos = 0, Y/T), Ch A (Scale = 5 V/Div, Ypos = 0, AC), Ch B (Scale = 5 V/Div, Ypos = 0, AC), Trigger (Pos edge, Level = 1 μV, Nor, A). Bring down the function generator enlargement and make sure that the following settings are selected: *Sine Wave*, Freq = 2 kHz, Ampl = 10 V, Offset = 0. Click the On-Off switch to run the simulation. Click "Pause" after ten screen displays on the oscilloscope. The red curve is plotting the voltage across the impedance of the R-C circuit (V), and the blue curve is plotting the current (I) because the voltage across the 1 kΩ resistor is proportional to the current (1 V is equivalent to 1 mA on the oscilloscope vertical axis). Draw the curve plots for the voltage (V) and current (I) in the space provided on the next page. Record the ac rms voltage (V) and current (I) readings on the ac voltmeter and ammeter.

V = __________ rms I = __________ rms

Step 8. Based on the curve plots in Step 7, determine the phase difference (θ) between the voltage and current.

Step 9. Based on the ac rms voltage (V) and current (I), calculate the magnitude of the impedance (Z) of the R-C circuit.

Step 10. Based on the value of the capacitance (C) and the sinusoidal frequency (f), calculate the capacitive reactance (X_C) of the capacitor.

Step 11. Based on the value of resistance R and the capacitive reactance (X_C) of capacitor C, calculate the expected magnitude of the impedance (Z) of the R-C circuit.

Question: How did your calculated impedance magnitude in Step 11 compare with the impedance from the measured ac rms voltage and current in Step 9?

Step 12. Based on the calculated capacitive reactance (X_C) and the value of resistance R, calculate the expected phase difference (θ) between the current and voltage sinusoidal functions.

Question: How did the calculated value for the phase difference in Step 12 compare with the measured phase difference between the current and voltage curve plots in Steps 7 and 8? Is the voltage leading or lagging the current? Is this what you expected?

Step 13. Pull down the File menu and open FIG21-3. Bring down the oscilloscope enlargement and make sure that the following settings are selected: Time base (Scale = 200 μs/Div, Xpos = 0, Y/T), Ch A (Scale 5 V/Div, Ypos = 0, AC), Ch B (Scale = 5 V/Div, Ypos = 0, AC). Trigger (Pos edge, Level = 1 μV, Nor, A). Bring down the function generator enlargement and make sure that the following settings are selected: *Sine Wave*, Freq = 1 kHz, Ampl = 10 V, Offset = 0. Click the On-Off switch to run the simulation. Click "Pause" after ten screen displays on the oscilloscope. The red curve is plotting the voltage across the impedance of the R-L-C circuit (V), and the blue curve is plotting the current (I) because the voltage across the 1 kΩ resistor is proportional to the current (1 V is equivalent to 1 mA on the oscilloscope vertical axis). Draw the curve plots for the voltage (V) and current (I) in the space provided on the next page. Record the ac rms input voltage (V), circuit current (I), inductor voltage (V_L), capacitor voltage (V_C), and resistor voltage (V_R) on the ac voltmeters and ammeter.

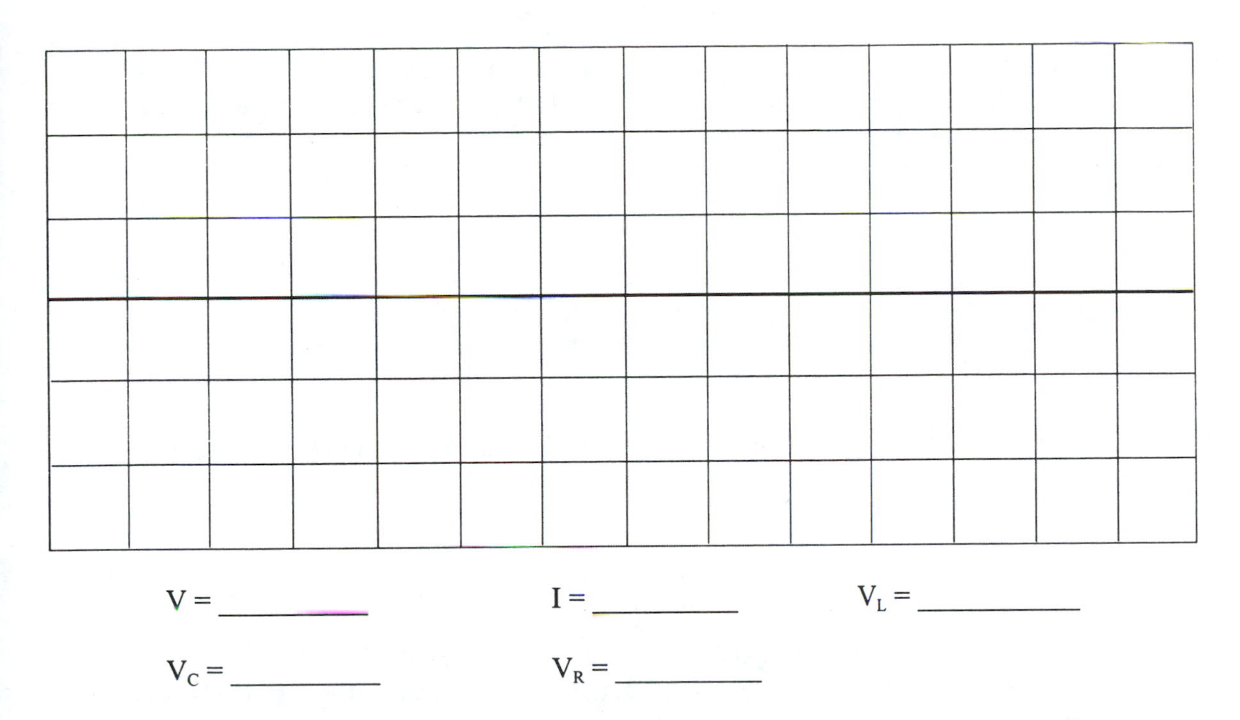

$V =$ __________ $I =$ __________ $V_L =$ __________

$V_C =$ __________ $V_R =$ __________

Step 14. Based on the curve plots in Step 13, determine the phase difference (θ) between the voltage and current.

Step 15. Based on the ac rms input voltage (V) and circuit current (I), calculate the magnitude of the impedance (Z) of the R-L-C circuit.

Step 16. Based on the value of the capacitance (C) and the sinusoidal frequency (f), calculate the capacitive reactance (X_C) of the capacitor.

Step 17. Based on the value of the inductance (L) and the sinusoidal frequency (f), calculate the inductive reactance (X_L) of the inductor.

Step 18. Based on the value of resistance R, the capacitive reactance (X_C) of capacitor C, and the inductive reactance (X_L) of inductor L, calculate the expected magnitude of the impedance (Z) of the R-L-C circuit.

Question: How did your calculated impedance magnitude in Step 18 compare with the impedance from the measured ac rms voltage and current in Step 15?

Step 19. Based on the calculated capacitive reactance (X_C), the calculated inductive reactance (X_L), and the value of resistance R, calculate the expected phase difference (θ) between the current and voltage sinusoidal functions.

Question: How did the calculated value for the phase difference in Step 19 compare with the measured phase difference between the current and voltage curve plots in Steps 13 and 14? Is the voltage leading or lagging the current? Is this what you expected? **Explain.**

Step 20. Based on the inductive reactance (X_L) calculated in Step 17 and the ac rms circuit current (I), calculate the expected ac rms voltage across the inductance (V_L).

Question: How did your calculated value for V_L in Step 20 compare with the measured value in Step 13?

Step 21. Based on the capacitive reactance (X_C) calculated in Step 16 and the ac rms circuit current (I), calculate the expected ac rms voltage across the capacitance (V_C).

Question: How did your calculated value for V_C in Step 21 compare with the measured value in Step 13?

Step 22. Based on the resistance (R) and the ac rms circuit current (I), calculate the expected ac rms voltage across the resistance (V_R).

Question: How did your calculated value for V_R in Step 22 compare with the measured value in Step 13?

Step 23. Add the **algebraic sum** of the ac rms voltages across the inductor, capacitor, and resistor.

Question: Does the algebraic sum of the voltages equal the ac rms voltage (V) applied across the total impedance? If not, why not? Did the sum satisfy Kirchhoff's voltage law?

Step 24. Add the **phasor sum** of the ac rms voltages across the inductor, capacitor, and resistor, taking the phase difference between the voltages into account. Draw the phasor diagram.

Question: Does the phasor sum of the voltages equal the ac rms voltage (V) applied across the total impedance? Did the sum satisfy Kirchhoff's voltage law?

Step 25. Change the frequency of the function generator to 2 kHz. Click the On-Off switch to run the simulation for ten screen displays on the oscilloscope. Record the ac rms circuit current (I) and the ac rms voltages across the inductor (V_L), across the capacitor (V_C), across the resistor (V_R), and across the total impedance (V) measured by the ac ammeter and ac voltmeters.

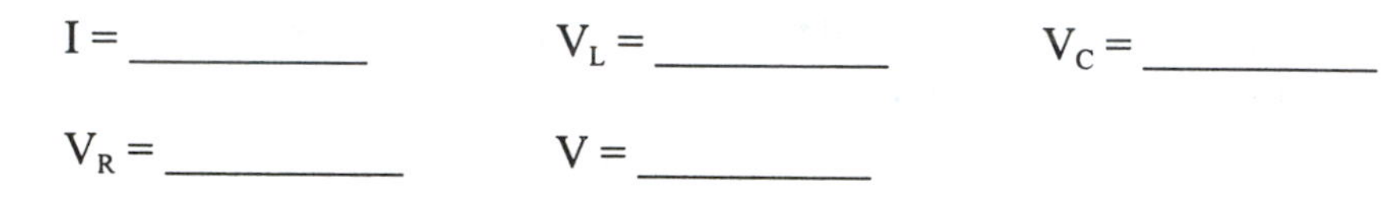

Step 26. Based on the new frequency (f) and the value of L, calculate the new inductive reactance (X_L).

Step 27. Based on the new frequency (f) and the value of C, calculate the new capacitive reactance (X_C).

Step 28. Based on the values of X_L, X_C, and R, calculate the new total circuit impedance (Z).

Step 29. Based on the applied rms voltage (V) across the total impedance and the impedance calculated in Step 28, calculate the expected ac rms circuit current (I).

Question: How did your calculated ac rms circuit current compare with the measured value in Step 25?

Step 30. Based on the inductive reactance (X_L) calculated in Step 26 and the ac rms circuit current (I), calculate the expected ac rms voltage across the inductance (V_L).

Question: How did your calculated value for V_L in Step 30 compare with the measured value in Step 25?

Step 31. Based on the capacitive reactance (X_C) calculated in Step 27 and the ac rms circuit current (I), calculate the expected ac rms voltage across the capacitance (V_C).

Question: How did your calculated value for V_C in Step 31 compare with the measured value in Step 25?

Step 32. Based on the resistance (R) and the ac rms circuit current (I), calculate the expected ac rms voltage across the resistance (V_R).

Question: How did your calculated value for V_R in Step 32 compare with the measured value in Step 25?

Step 33. Add the **phasor sum** of the ac rms voltages across the inductor, capacitor, and resistor, taking the phase difference between the voltages into account. Draw the phasor diagram.

Question: Does the phasor sum of the voltages equal the ac rms voltage (V) applied across the total impedance? Did the sum satisfy Kirchhoff's voltage law?

Troubleshooting Problems

Note: Exercises 5 and 6 are challenging problems for more advanced students.

1. Pull down the File menu and open FIG21-4. Click the On-Off switch to run the simulation. Based on the oscilloscope curve plots, which component is shorted (R, L, or C)? **Explain why**.

2. Pull down the File menu and open FIG21-5. Click the On-Off switch to run the simulation. Based on the oscilloscope curve plots, which component is shorted (R, L, or C)? **Explain why**.

3. Pull down the File menu and open FIG21-6. Click the On-Off switch to run the simulation. The red curve is plotting the voltage across X and the blue curve is plotting the current in X. Based on the oscilloscope curve plots, determine the component in X (R, L, or C). **Explain why**.

4. Pull down the File menu and open FIG21-7. Click the On-Off switch to run the simulation. The red curve is plotting the voltage across X and the blue curve is plotting the current in X. Based on the oscilloscope curve plots, determine the component in X (R, L, or C). **Explain why**.

5. Pull down the File menu and open FIG21-8. Click the On-Off switch to run the simulation for ten screen displays, then click "pause." Based on the oscilloscope curve plots, determine the value of L and R.

L = __________ R = __________

6. Pull down the File menu and open FIG21-9. Click the On-Off switch to run the simulation until the meter readings are stable. Based on the meter readings, determine the value of L and C.

L = __________ C = __________

EXPERIMENT

22

Name________________________________

Date________________________________

Admittance of Parallel AC Circuits

Objectives:

1. Measure the admittance of a parallel R-L circuit and the phase between the ac voltage and current and compare your measured values with the calculated values.
2. Measure the admittance of a parallel R-C circuit and the phase between the ac voltage and current and compare your measured values with the calculated values.
3. Measure the admittance of a parallel R-L-C circuit and the phase between the ac voltage and current and compare your measured values with the calculated values.
4. Calculate the expected ac rms current in each branch in a parallel R-L-C circuit and compare your calculated currents with the measured values.
5. Demonstrate how Kirchhoff's current law applies to a parallel ac admittance using phasors.
6. Demonstrate the effect of frequency changes on the ac rms currents in a parallel R-L-C circuit.

Materials:

One dual-trace oscilloscope
One function generator
Four 0–20 mA ac milliammeters
One 0–10 V ac voltmeter
One 0.1 μF capacitor
One 100 mH inductor
Resistors—1 Ω, 1 kΩ

Theory:

Complete Experiments 19 and 20 on inductive and capacitive reactance before attempting this experiment. Also review the Theory section of Experiment 21.

When an ac sinusoidal voltage is applied across a parallel R-L, R-C, or R-L-C circuit, as shown in Figures 22-1, 22-2, and 22-3, there is an opposition to the ac current entering the parallel circuit called **impedance (Z)**, and its unit of measurement is the ohm (Ω). The current in each element is also sinusoidal and has the same frequency as the applied voltage. However, the ac current in the inductor will lag the applied ac voltage by 90 degrees, the ac current in the capacitor will lead the applied ac voltage by 90 degrees, and the ac current in the resistor will be in-phase with the applied ac voltage. This will cause a **phase difference** (θ) between the ac voltage applied to the parallel circuit and the ac current entering the parallel circuit. This phase difference can be **between 0 degrees and 90 degrees**, depending on the relationship between the total **reactance** and total **resistance** in the circuit.

When dealing with parallel ac circuits, the inverse of impedance, called **admittance (Y)**, is easier to use.

Therefore,

$$Y = \frac{1}{Z}$$

The unit of measurement for admittance is the **siemen (S)**.

As in a dc resistive circuit, **Ohm's law** determines the relationship between the ac rms voltage applied to the parallel circuit (V) and the ac rms current entering the parallel circuit (I). In an ac circuit, the impedance (Z) of the circuit replaces the resistance of a resistor in the Ohm's law equation. Therefore,

$$V = Z\,I$$

where Z is the circuit impedance in ohms (Ω). Because admittance is the inverse of impedance, **Ohm's law for admittance** is

$$V = \frac{1}{Y} I$$

or

$$I = Y\,V$$

Because the rms voltage and current are both equal to the peak value times 0.707, Ohm's law can also be applied to the peak ac voltage (V_P) and peak ac current (I_P) as follows:

$$I_P = Y\,V_P$$

Because the ac current in the parallel reactance (I_X) is 90 degrees out-of-phase with the applied ac voltage (V), and the ac current in the parallel resistance (I_R) is in phase with the applied ac voltage, they must be treated as **phasors**. Therefore, I_R and I_X must be added as if they are separated by 90 degrees. From right angle trigonometry (Pythagorean theorem), **Kirchhoff's current law**, and Ohm's law, the ac rms current entering the parallel circuit (I) can be represented by

$$I = \sqrt{(I_R)^2 + (I_X)^2} = \sqrt{\left(\frac{V}{R}\right)^2 + \left(\frac{V}{X}\right)^2} = \sqrt{(GV)^2 + (BV)^2} = V\sqrt{G^2 + B^2} = VY$$

Based on this equation, the parallel circuit admittance (Y) can be determined from

$$Y = \sqrt{G^2 + B^2}$$

where $G = 1/R$ is the circuit **conductance** in siemens, and $B = 1/X$ is the circuit **susceptance** in siemens.

From right angle trigonometry (Pythagorean theorem), the **phase difference** between the ac rms voltage applied across the parallel circuit and the ac rms current entering the parallel circuit can be calculated from

$$\theta = \arctan\left(\frac{BV}{GV}\right) = \arctan\left(\frac{B}{G}\right)$$

where θ is the phase difference in radians or degrees. As can be seen from the equation, the larger the susceptance (B) compared to the conductance (G), the larger the phase difference (θ) between the ac rms voltage applied across the parallel circuit and the ac rms current entering the parallel circuit.

The inductive susceptance (B_L), capacitive susceptance (B_C), and conductance (G) can be used to calculate I_L, I_C, and I_R from

$$I_L = VB_L$$

for an inductor,

$$I_C = VB_C$$

for a capacitor, and

$$I_R = VG$$

for a resistor.

When applying Kirchhoff's current law to a parallel ac admittance, the currents must be added using **phasor addition** because they are out-of-phase with each other. Both the magnitude and the phase of each current must be taken into account.

The phase difference (θ) between two periodic functions (such as sine functions) can be measured by measuring the time difference (t) between the two curve plots and the time period for one cycle (T) of the curve plots. Because the ratio of the time difference (t) divided by the time period (T) for one cycle is equal to the ratio of the phase difference (θ), in degrees, divided by the number of degrees of phase for one cycle (360 degrees), the phase difference can be calculated from

$$\frac{\theta}{360^\circ} = \frac{t}{T}$$

Therefore,

$$\theta = \left(\frac{t}{T}\right)360^\circ$$

The admittance (Y) of the parallel R-L ac circuit in Figure 22-1 is the phasor sum of the conductance (G), in siemens, and the inductive susceptance (B_L), in siemens. Therefore, the magnitude of the admittance (Y) can be found from

$$Y = \sqrt{G^2 + {B_L}^2}$$

and the phase difference (θ) between the ac rms voltage across the parallel circuit (V) and the ac rms current entering the parallel circuit (I) can be found from

$$\theta = -\arctan\left(\frac{B_L}{G}\right)$$

Note that for admittances, when the current (I) lags the voltage (V), the phase difference (θ) is negative.

The **conductance** (G) is the inverse of the resistance (R). Therefore,

$$G = \frac{1}{R}$$

The **inductive susceptance** (B_L) is the inverse of the inductive reactance (X_L). Therefore,

$$B_L = \frac{1}{X_L} = \frac{1}{2\pi fL}$$

The admittance (Y) of the parallel R-C ac circuit in Figure 22-2 is the phasor sum of the conductance (G), in siemens, and the capacitive susceptance (B_C), in siemens. Therefore, the magnitude of the admittance (Y) can be found from

$$Y = \sqrt{G^2 + B_C{}^2}$$

and the phase difference (θ) between the ac rms voltage across the parallel circuit (V) and the ac rms current entering the parallel circuit (I) can be found from

$$\theta = \arctan\left(\frac{B_C}{G}\right)$$

The **capacitive susceptance** (B_C) is the inverse of the capacitive reactance (X_C). Therefore,

$$B_C = \frac{1}{X_C} = 2\pi fC$$

The admittance (Y) of the parallel R-L-C ac circuit in Figure 22-3 is the phasor sum of the conductance (G), in siemens, and the total susceptance (B), in siemens. The total susceptance (B) is equal to the phasor sum of the inductive susceptance (B_L) and the capacitive susceptance (B_C). Because the inductive susceptance and capacitive susceptance phasors are 180 degrees out of phase, the total susceptance (B) is

$$B = B_C - B_L$$

Therefore, the magnitude of the admittance (Y) for the parallel R-L-C circuit can be found from

$$Y = \sqrt{G^2 + B^2}$$

and the phase difference (θ) between the ac rms voltage across the parallel circuit (V) and the ac rms current entering the parallel circuit (I) can be found from

$$\theta = \arctan\left(\frac{B}{G}\right)$$

The inductive susceptance (B_L) and the capacitive susceptance (B_C) are a function of the ac sinusoidal frequency. In a parallel R-L-C ac circuit, there is only one frequency at which they are equal. At this frequency, the total susceptance is zero ($B = B_C - B_L = 0$) and the circuit admittance is $Y = G$ and is at the minimum value, making the impedance (Z) be at its maximum value.

In the parallel R-L-C circuit in Figure 22-3, the ac rms current in the resistance (I_R) is in-phase with the ac rms voltage across the parallel circuit (V), the ac rms current in the inductor (I_L) lags the voltage (V) by 90 degrees, and the ac rms current in the capacitor (I_C) leads the voltage (V) by 90 degrees. The phasor diagram representing the current phasors is shown below.

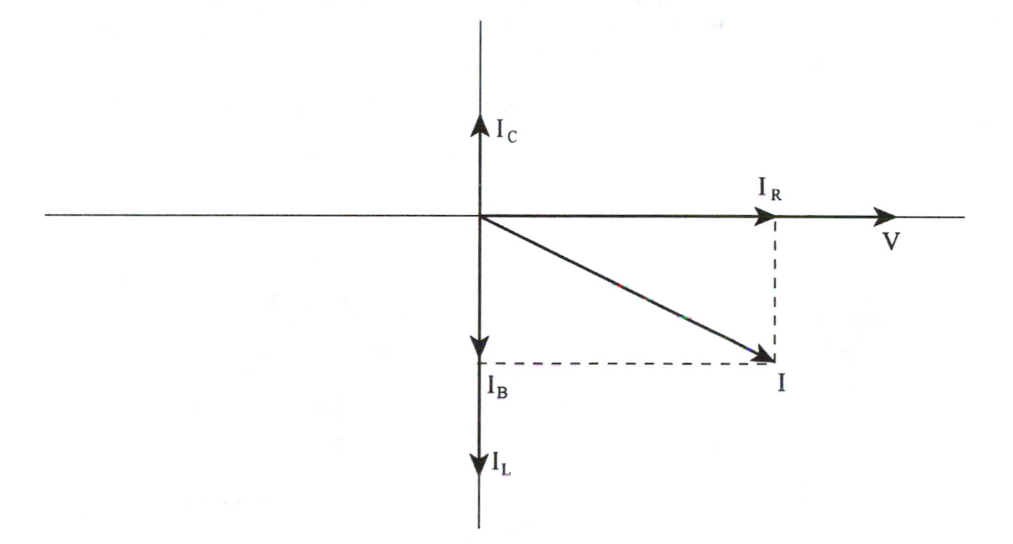

Because phasors I_L and I_C are 180 degrees out-of-phase, the total current in the inductor and capacitor (I_B) is equal to the difference between I_C and I_L. Therefore,

$$I_B = I_C - I_L$$

Notice that the phasor sum of I_R and I_B is equal to I. From the triangle,

$$I = \sqrt{I_R^2 + I_B^2}$$

Note: If you do the experiment in a real laboratory environment (hardwired), see the end of the Theory section of Experiment 19 regarding the inductor resistance (R_L).

Figure 22-1 Admittance—Parallel R-L Circuit

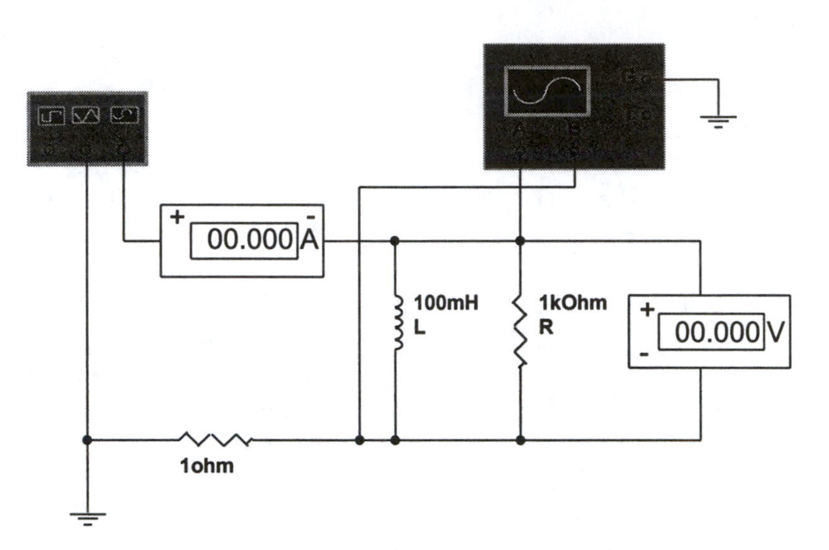

Figure 22-2 Admittance—Parallel R-C Circuit

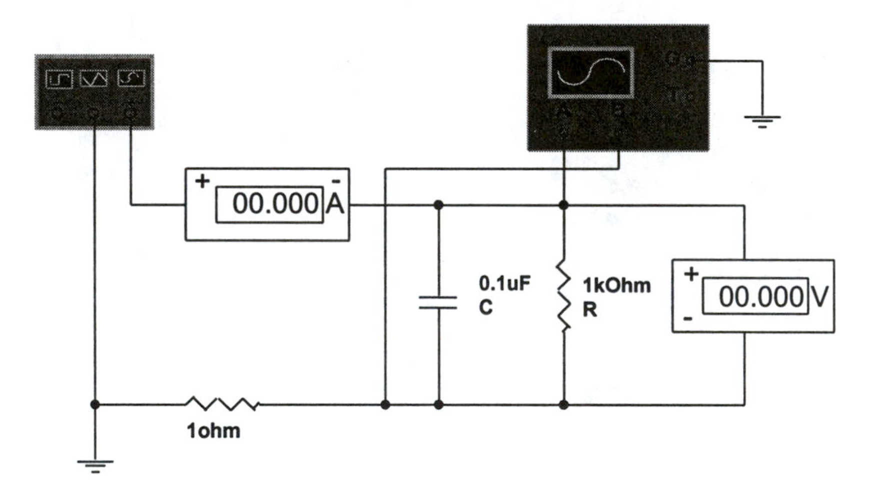

Figure 22-3 Admittance—Parallel R-L-C Circuit

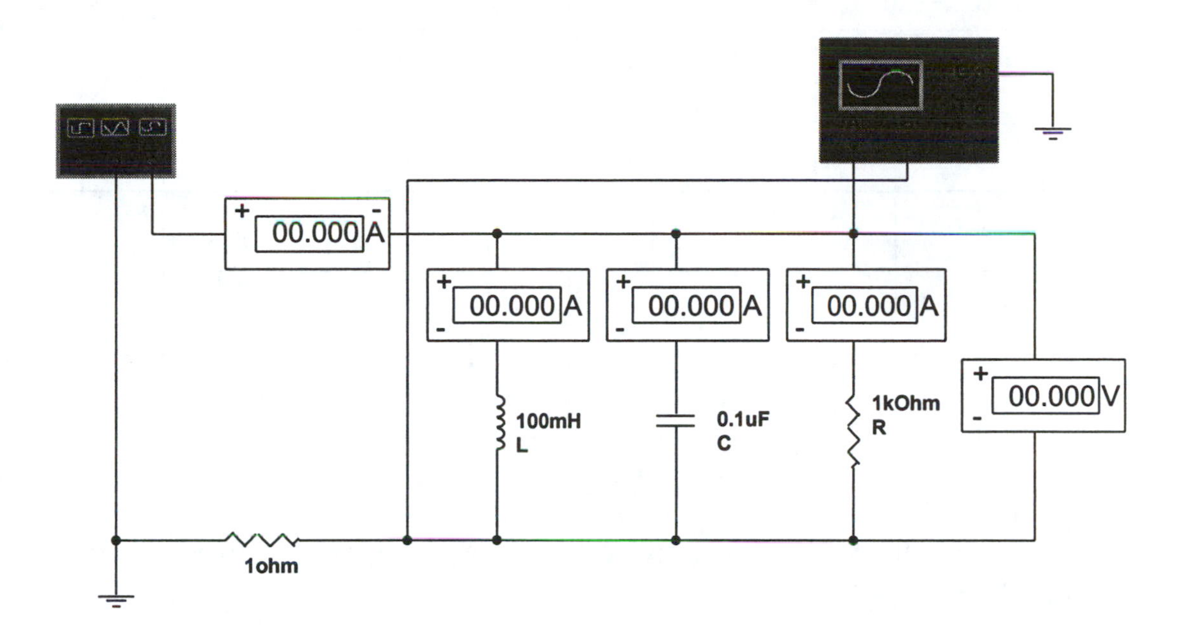

Procedure:

Step 1. Pull down the File menu and open FIG22-1. Bring down the oscilloscope enlargement and make sure that the following settings are selected: Time base (Scale = 200 μs/Div, Xpos = 0, Y/T), Ch A (Scale = 5 V/Div, Ypos = 0, AC), Ch B (Scale = 10 mV/Div, Ypos = 0, AC), Trigger (Pos edge, Level = 1 μV, Nor A). Bring down the function generator enlargement and make sure that the following settings are selected: *Sine Wave*, Freq = 1 kHz, Ampl = 10 V, Offset = 0. Click the On-Off switch to run the simulation. Click "Pause" after two screen displays on the oscilloscope. The red curve on the oscilloscope is plotting the voltage across the parallel R-L circuit (V), and the blue curve is plotting the current entering the parallel R-L circuit (I) because the voltage across the 1 Ω resistor is proportional to the current (1 mV is equivalent to 1 mA on the oscilloscope vertical axis). Draw the curve plots for the voltage (V) and current (I) in the space provided on the following page. Record the ac rms voltage (V) and current (I) readings on the ac voltmeter and ammeter.

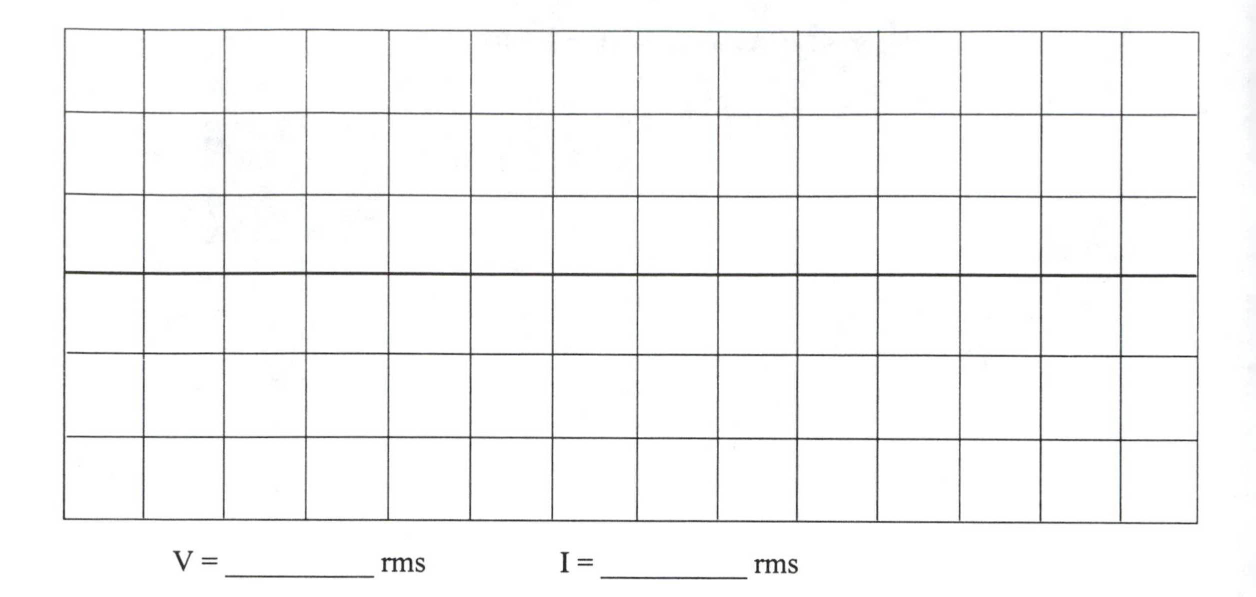

V = __________ rms I = __________ rms

Step 2. Based on the curve plots in Step 1, calculate the phase difference (θ) between the voltage and current.

Step 3. Based on the ac rms voltage (V) and current (I), calculate the magnitude of the admittance (Y) of the parallel R-L circuit. From the value of the admittance (Y), calculate the impedance (Z) of the parallel R-L circuit.

Step 4. Based on the value of the inductance (L) and the sinusoidal frequency (f), calculate the inductive susceptance (B_L) of the inductor.

Step 5. Based on the resistance of resistor R, calculate the conductance (G) of the resistor.

Step 6. Based on the conductance (G) of resistor R and the inductive susceptance (B_L) of inductor L, calculate the expected magnitude of the admittance (Y) of the parallel R-L circuit. From the value of the admittance (Y), calculate the impedance (Z) of the parallel R-L circuit.

Question: How did your calculated admittance and impedance magnitudes in Step 6 compare with the admittance and impedance calculated from the measured ac rms voltage and current in Step 3?

Step 7. Based on the inductive susceptance (B_L) and the conductance (G), calculate the expected phase difference (θ) between the current and voltage sinusoidal functions.

Question: How did the calculated value for the phase difference in Step 7 compare with the measured phase difference between the current and voltage curve plots in Steps 1 and 2? Is the current leading or lagging the voltage? Is this what you expected?

Step 8. Pull down the File menu and open FIG22-2. Bring down the oscilloscope enlargement and make sure that the following settings are selected: Time base (Scale = 100 μs/Div, Xpos = 0, Y/T), Ch A (Scale = 5 V/Div, Ypos = 0, AC), Ch B (Scale = 10 mV/Div, Ypos = 0, AC), Trigger (Pos edge, Level = 1 μV, Sing, A). Bring down the function generator enlargement and make sure that the following settings are selected: *Sine Wave*, Freq = 2 kHz, Ampl = 10 V, Offset = 0. Click the On-Off switch to run the simulation. The red curve on the oscilloscope is plotting the voltage across the parallel R-C circuit (V), and the blue curve is plotting the current entering the parallel R-C circuit (I) because the voltage across the 1 Ω resistor is proportional to the current (1 mV is equivalent to 1 mA on the oscilloscope vertical axis). Draw the curve plots for the voltage (V) and current (I) in the space provided. Record the ac rms voltage (V) and current (I) readings on the ac voltmeter and ammeter.

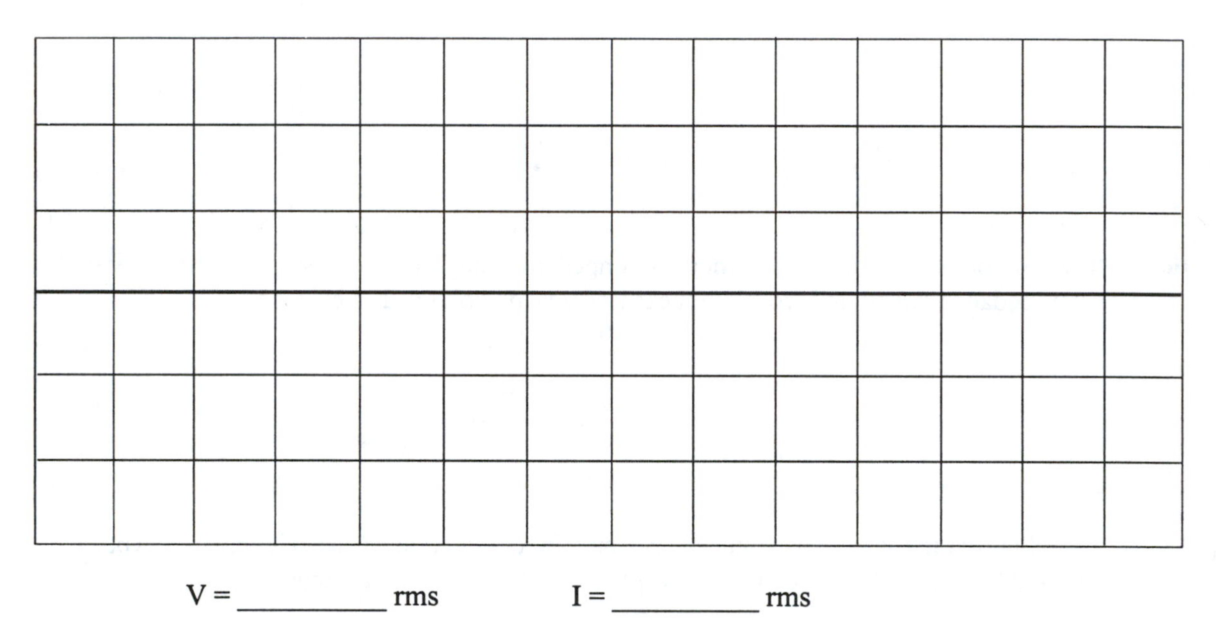

V = __________ rms   I = __________ rms

Step 9. Based on the curve plots in Step 8, calculate the phase difference (θ) between the voltage and current.

Step 10. Based on the ac rms voltage (V) and current (I), calculate the magnitude of the admittance (Y) of the parallel R-C circuit. From the value of the admittance (Y), calculate the impedance (Z) of the parallel R-C circuit.

Step 11. Based on the value of the capacitance (C) and the sinusoidal frequency (f), calculate the capacitive susceptance (B_C) of the capacitor.

Step 12. Based on the resistance of resistor R, calculate the conductance (G) of the resistor.

Step 13. Based on the conductance (G) of resistor R and the capacitive susceptance (B_C) of capacitor C, calculate the expected magnitude of the admittance (Y) of the parallel R-C circuit. From the value of the admittance (Y), calculate the impedance (Z) of the parallel R-C circuit.

Question: How did your calculated admittance and impedance magnitudes in Step 13 compare with the admittance and impedance calculated from the measured ac rms voltage and current in Step 10?

Step 14. Based on the capacitive susceptance (B_C) and the conductance (G), calculate the expected phase difference (θ) between the current and voltage sinusoidal functions.

Question: How did the calculated value for the phase difference in Step 14 compare with the measured phase difference between the current and voltage curve plots in Steps 8 and 9? Is the current leading or lagging the voltage? Is this what you expected?

Step 15. Pull down the File menu and open FIG22-3. Bring down the oscilloscope enlargement and make sure that the following settings are selected: Time base (Scale = 200 μs/Div, Xpos = 0, Y/T), Ch A (Scale = 5 V/Div, Ypos = 0, AC), Ch B (Scale = 10 mV/Div, Ypos = 0, AC), Trigger (Pos edge, Level = 1 μV, Sing, A). Bring down the function generator enlargement and make sure that the following settings are selected: *Sine Wave*, Freq = 1 kHz, Ampl = 10 V, Offset = 0. Click the On-Off switch to run the simulation. Click "Pause" after fifteen screen displays on the oscilloscope. The red curve is plotting the voltage across the parallel R-L-C circuit (V), and the blue curve is plotting the current entering the parallel R-L-C circuit (I) because the voltage across the 1 Ω resistor is proportional to the current (1 mV is equivalent to 1 mA on the oscilloscope vertical axis). Draw the curve plots for the voltage (V) and current (I) in the space provided. Record the ac rms voltage (V) across the R-L-C circuit, the circuit current (I), the inductor current (I_L), the capacitor current (I_C), and the resistor current (I_R) on the ac voltmeter and ammeters.

V = __________ I = __________ I_L = __________

I_C = __________ I_R = __________

Step 16. Based on the curve plots in Step 15, calculate the phase difference (θ) between the voltage and current.

Step 17. Based on the ac rms voltage (V) and circuit current (I), calculate the magnitude of the admittance (Y) of the parallel R-L-C circuit. From the value of the admittance (Y), calculate the impedance (Z) of the parallel R-L-C circuit.

Step 18. Based on the value of the capacitance (C) and the sinusoidal frequency (f), calculate the capacitive susceptance (B_C) of the capacitor.

Step 19. Based on the value of the inductance (L) and the sinusoidal frequency (f), calculate the inductive susceptance (B_L) of the inductor.

Step 20. Based on the resistance of resistor R, calculate the conductance (G) of the resistor.

Step 21. Based on the conductance (G) of resistor R, the capacitive susceptance (B_C) of capacitor C, and the inductive susceptance (B_L) of inductor L, calculate the expected magnitude of the admittance (Y) of the R-L-C circuit. From the value of the admittance (Y), calculate the impedance (Z) of the R-L-C circuit.

Question: How did your calculated admittance and impedance magnitudes in Step 21 compare with the admittance and impedance calculated from the measured ac rms voltage and current in Step 17?

Step 22. Based on the calculated capacitive susceptance (B_C), the calculated inductive susceptance (B_L), and the conductance (G), calculate the expected phase difference (θ) between the current and voltage sinusoidal functions.

Question: How did the calculated value for the phase difference in Step 22 compare with the measured phase difference between the current and voltage curve plots in Steps 15 and 16? Is the current leading or lagging the voltage? Is this what you expected? **Explain.**

Step 23. Based on the inductive susceptance (B_L) calculated in Step 19 and the ac rms voltage across the parallel circuit (V), calculate the expected ac rms current in the inductance (I_L).

Question: How did your calculated value of I_L in Step 23 compare with the measured value in Step 15?

Step 24. Based on the capacitive susceptance (B_C) calculated in Step 18 and the ac rms voltage across the parallel circuit (V), calculate the expected ac rms current in the capacitance (I_C).

Question: How did your calculated value of I_C in Step 24 compare with the measured value in Step 15?

Step 25. Based on the conductance (G) calculated in Step 20 and the ac rms voltage (V) across the parallel circuit, calculate the expected ac rms current in the resistance (I_R).

Question: How did your calculated value of I_R in Step 25 compare with the measured value in Step 15?

Step 26. Add the **algebraic sum** of the ac rms currents in the inductor, capacitor, and resistor.

Question: Does the algebraic sum of the currents equal the ac rms current entering the parallel circuit? If not, why not? Did the sum satisfy Kirchhoff's current law?

Step 27. Add the **phasor sum** of the ac rms currents in the inductor, capacitor, and resistor, taking the phase difference between the currents into account. Draw the phasor diagram.

Question: Does the phasor sum of the currents equal the ac rms current entering the parallel circuit? Did the sum satisfy Kirchhoff's current law?

Step 28. Change the frequency of the function generator to 2 kHz. Click the On-Off switch to run the simulation for ten screen displays on the oscilloscope. Record the ac rms voltage across the parallel circuit (V), the total current entering the parallel circuit (I), and the ac rms current in the inductor (I_L), the capacitor (I_C), and the resistor (I_R) measured by the ac ammeters and ac voltmeter.

V = __________ I = __________ I_L = __________

I_C = __________ I_R = __________

Step 29. Based on the new frequency (f) and the value of L, calculate the new inductive susceptance (B_L).

Step 30. Based on the new frequency (f) and the value of C, calculate the new capacitive susceptance (B_C).

Step 31. Based on the new values of B_L and B_C, and the conductance G calculated in Step 20, calculate the new total circuit admittance (Y).

Step 32. Based on the applied ac rms voltage (V) across the total admittance and the new admittance (Y) calculated in Step 31, calculate the expected ac rms circuit current (I) entering the parallel circuit.

Question: How did your calculated ac rms current (I) compare with the measured value in Step 28?

Step 33. Based on the new inductive susceptance (B_L) calculated in Step 29 and the ac rms voltage across the parallel circuit (V), calculate the expected ac rms current in the inductance (I_L).

Question: How did your calculated value of I_L in Step 33 compare with the measured value in Step 28?

Step 34. Based on the new capacitive susceptance (B_C) calculated in Step 30 and the ac rms voltage across the parallel circuit (V), calculate the expected ac rms current in the capacitance (I_C).

Question: How did your calculated value of I_C in Step 34 compare with the measured value in Step 28?

Step 35. Based on the conductance (G) calculated in Step 20 and the ac rms voltage (V) across the parallel circuit, calculate the expected ac rms current in the resistance (I_R).

Question: How did your calculated value of I_R in Step 35 compare with the measured value in Step 28?

Step 36. Add the **phasor sum** of the ac rms currents in the inductor, capacitor, and resistor, taking the phase difference between the currents into account. Draw the phasor diagram.

Question: Does the phasor sum of the currents equal the ac rms current entering the parallel circuit? Did the sum satisfy Kirchhoff's current law?

Troubleshooting Problems

Note: Exercises 3 and 4 are challenging problems for more advanced students.

1. Pull down the File menu and open FIG22-4. Bring down the oscilloscope enlargement and click the On-Off switch to run the simulation. The blue curve is plotting the current entering the parallel R-L-C circuit and the red curve is plotting the voltage across the R-L-C circuit. Based on the oscilloscope curve plots, which component is open (R, L, or C)? **Explain**.

2. Pull down the File menu and open FIG22-5. Bring down the oscilloscope enlargement and click the On-Off switch to run the simulation. The blue curve is plotting the current entering the parallel R-L-C circuit and the red curve is plotting the voltage across the R-L-C circuit. Based on the oscilloscope curve plots, which component is open (R, L, or C)? **Explain**.

3. Pull down the File menu and open FIG22-6. Click the On-Off switch to run the simulation. Click "Pause" after thirty screen displays on the oscilloscope. Based on the oscilloscope curve plots, determine the value of R and C.

R = __________ C = __________

4. Pull down the File menu and open FIG22-7. Click the On-Off switch to run the simulation. Based on the meter readings, determine the value of L, C, and R.

L = __________ C = __________ R = __________

EXPERIMENT

23

Name______________________________

Date______________________________

Power and Power Factor in AC Circuits

Objectives:

1. Determine the real, reactive, and apparent power and the power factor for a series R-C circuit.
2. Determine the real, reactive, and apparent power and the power factor for a series R-L circuit.
3. Determine the real, reactive, and apparent power and the power factor for a series R-L-C circuit.
4. Determine the capacitance needed for power factor correction of a series R-L circuit.

Materials:

One dual-trace oscilloscope
One 12 V ac transformer
Two 0–20 mA ac milliammeters
Three 0–20 V ac voltmeters
One 1 H inductor
One 2 μF capacitor
Resistors—1 Ω, 500 Ω, 1 kΩ

Theory:

For time varying voltages and currents, the power delivered to a load is also time varying. This time-varying power is referred to as **instantaneous power**. This instantaneous power can be positive or negative at any given time. The real (actual) power delivered to the load is the average value of the instantaneous power.

For ac sinusoidal voltages and currents, the **real power (P)**, in watts, dissipated in an ac R-L, R-C, or R-L-C circuit is dissipated in the resistance only. There is no real (actual) power dissipation in a reactive element such as an inductor or capacitor. In a reactive element, energy is stored during one-half the ac sinusoidal cycle and released during the other half of the ac sinusoidal cycle. The power in a reactive element is called **reactive power (Q)**, and is measured in **vars**. The real power (P) dissipated in an ac load can be calculated from

$$P = I^2R$$

where R is the load resistance and I is the ac rms current. The reactive power in an ac load can be calculated from

$$Q = I^2X$$

where X is the load reactance and I is the ac rms current.

When an ac load has an ac rms voltage (V) across it and an ac rms current (I), the **apparent power (S)** is the product of the voltage and current, in volt-amperes (VA). Therefore, the apparent power can be calculated from

$$S = V\,I$$

If the load has both resistance and reactance, apparent power represents neither real power nor reactive power. It is called apparent power because it appears to represent the real power.

A power triangle can be drawn from the values of the real, reactive, and apparent power. The real power is the length of the horizontal axis, the reactive power is the length of the vertical axis, and the apparent power is the length of the hypotenuse of the triangle, as shown below.

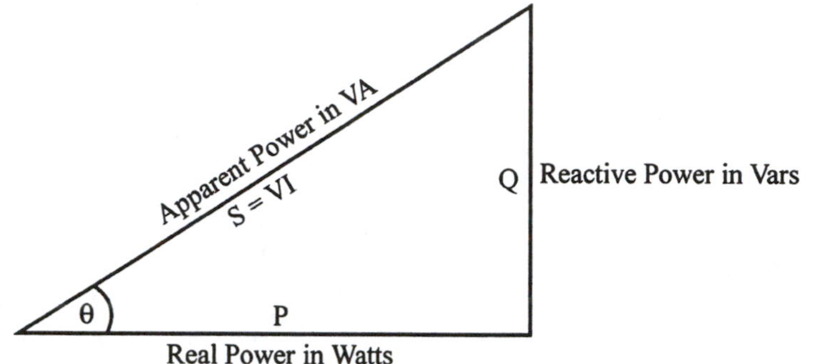

From the geometry of this triangle, S can be calculated from

$$S = \sqrt{P^2 + Q^2}$$

and **cos θ** is defined as the **power factor (pf).** The power factor (pf) is the ratio of the real power (P) divided by the apparent power (S) and can be found from

$$pf = \cos\theta = \frac{P}{S} = \frac{P}{VI}$$

where θ is the phase difference between the voltage across the load (V) and the load current (I). The power factor is considered to be lagging when the load current lags the load voltage (inductive load) and leading when the load current leads the load voltage (capacitive load).

The real power (P) in an ac circuit can also be found from the apparent power by multiplying the apparent power by the power factor. Therefore,

$$P = VI\cos\theta$$

The real power (P), in watts, dissipated in the circuits in Figures 23-1, 23-2, and 23-3 can be calculated from the ac rms resistor current (I) and the resistor value (R) using the equation

$$P = I^2R$$

The reactive power (Q) in the R-C circuit in Figure 23-1 can be calculated from

$$Q = V_C I = I^2 X_C$$

where V_C is the ac rms voltage across the capacitor, I is the ac rms capacitor current, and X_C is the capacitive reactance.

The reactive power (Q) in the R-L circuit in Figure 23-2 can be calculated from

$$Q = V_L I = I^2 X_L$$

where V_L is the ac rms voltage across the inductor, I is the ac rms inductor current, and X_L is the inductive reactance.

The reactive power (Q) in the R-L-C circuit in Figure 23-3 can be calculated from

$$Q = V_X I = I^2 X$$

where $V_X = V_C - V_L$ is equal to the ac rms voltage across the total reactance, I is the current in the reactance, and $X = X_C - X_L$ is the total reactance. The ac rms voltage across the total reactance is equal to the difference between the capacitor voltage (V_C) and the inductor voltage (V_L) because these voltages are 180 degrees out-of-phase with each other.

Power factor correction is normally required for an inductive load because most ac motors are inductive (R-L circuit). Because a unity power factor load requires less current, it is advantageous to bring the power factor of an inductive load, such as an ac motor, as close to unity as possible. This will make the real power (P) close to being equal to the apparent power (VI). To find the capacitance value needed to correct the power factor in the inductive load in Figure 23-4, first you need to determine the reactive power of the original R-L circuit. This can be accomplished by drawing the power triangle and solving for the reactive power (Q) from the triangle. The power triangle can be drawn from the real power (P), the apparent power (VI), and the power factor angle (θ).

Once the reactive power (Q) for the original R-L circuit has been determined, the capacitive reactance (X_C) needed for power factor correction can be found from

$$Q = \frac{V^2}{X_C}$$

where V is the voltage across the R-L circuit. Therefore,

$$X_C = \frac{V^2}{Q}$$

Once X_C has been calculated, the capacitance required for power factor correction can be found from

$$X_C = \frac{1}{2\pi fC}$$

Therefore,

$$C = \frac{1}{2\pi fX_C}$$

If the correct capacitor has been chosen and placed across the R-L load (motor), the power factor will be close to unity (voltage V in-phase with current I). This will make the real power nearly equal to the apparent power.

If pf = cos θ = P/S = 1, then P = S.

> *Note:* If you do the experiment in a real laboratory environment (hardwired), see the end of the Theory section of Experiment 19 regarding the inductor resistance (R_L).

Figure 23-1 AC Power in R-C Circuits

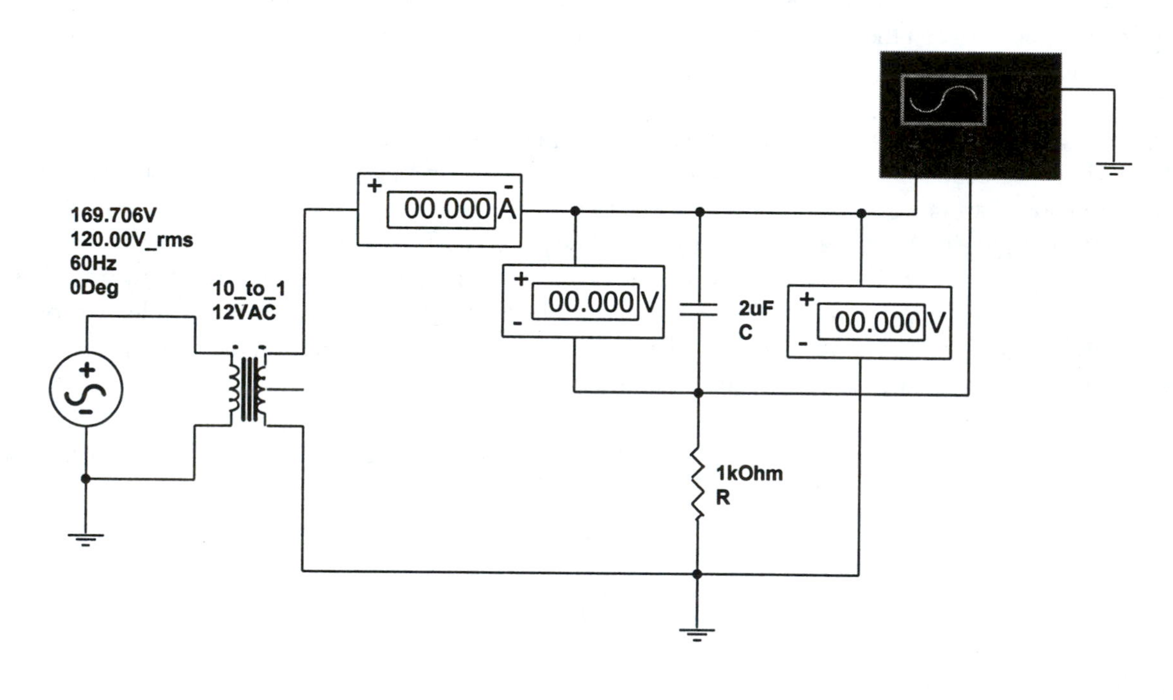

Figure 23-2 AC Power in R-L Circuits

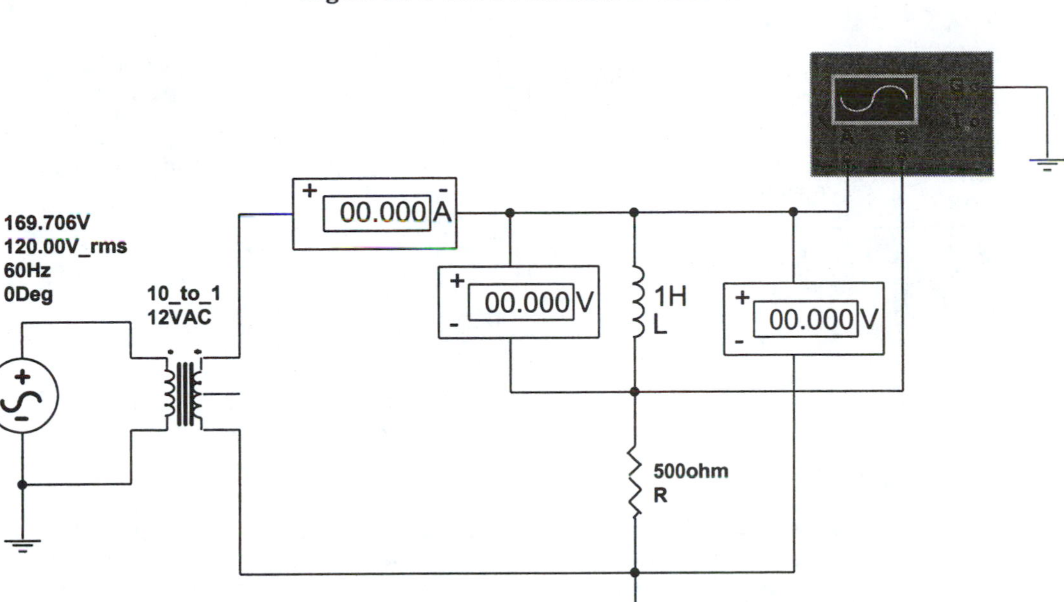

Figure 23-3 AC Power in R-L-C Circuits

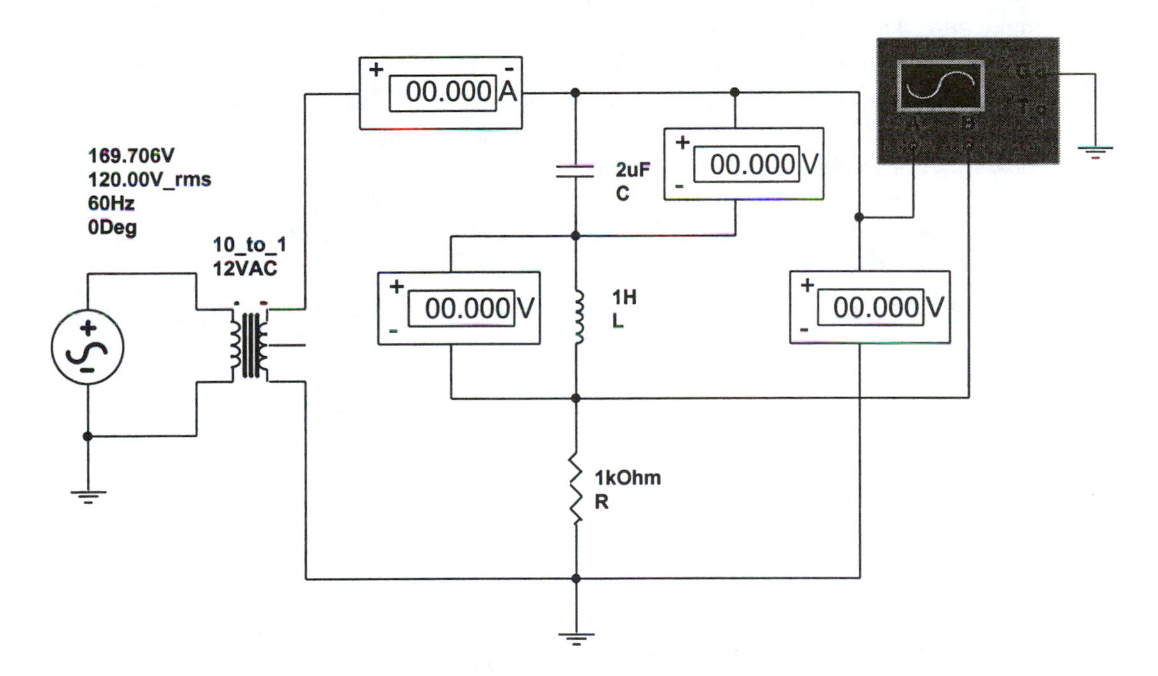

Figure 23-4 Power Factor Correction

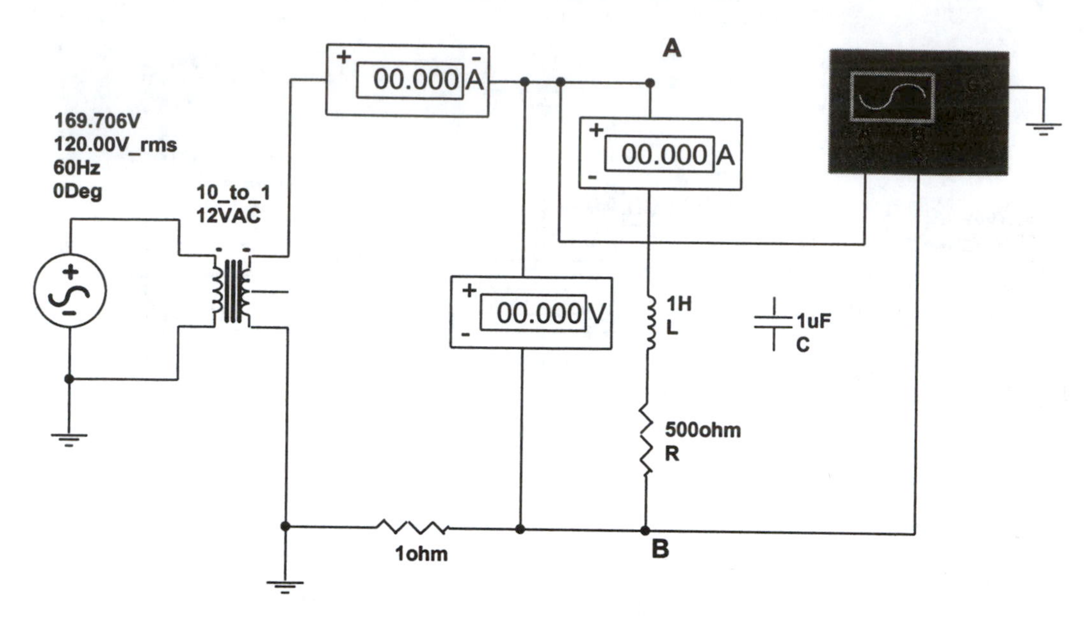

Procedure:

Step 1. Pull down the File menu and open FIG23-1. Bring down the oscilloscope enlargement and make sure that the following settings are selected: Time base (Scale = 2 ms/Div, Xpos = 0, Y/T), Ch A (Scale = 10 V/Div, Ypos = 0, AC), Ch B (Scale = 5 V/Div, Ypos = 0, AC), Trigger (Pos edge, Level = 0, Nor, A). Click the On-Off switch and run the simulation to completion (Tran = 0.210 s). Record the total ac rms current (I), the ac rms voltage across the capacitor (V_C), and the total ac rms voltage across the R-C network (V).

I = __________ rms V_C = __________ rms V = __________ rms

Step 2. Based on the readings in Step 1, calculate the real ac power (P) in the R-C circuit.

Step 3. Based on the readings in Step 1, calculate the reactive power (Q) in the R-C circuit.

Step 4. Based on the readings in Step 1, calculate the apparent power (S) in the R-C circuit.

Step 5. Based on the calculated real power (P), the reactive power (Q), and the apparent power (S) in Steps 2–4, draw the power triangle and determine the power factor (pf) for the R-C network.

Step 6. The oscilloscope is plotting voltage V (red curve plot) and current I (blue curve plot). Using the oscilloscope curve plots, determine the phase (θ) between voltage V and current I. (See the Theory section in Experiment 21.) Based on the value of the phase (θ), calculate the power factor (pf).

Question: How did your measured power factor in Step 6 compare with the value determined from the power triangle in Step 5? Is it lagging or leading? **Explain why**.

Step 7. Pull down the File menu and open FIG23-2. Bring down the oscilloscope enlargement and make sure that the following settings are selected: Time base (Scale = 2 ms/Div, Xpos = 0, Y/T), Ch A (Scale = 10 V/Div, Ypos = 0, AC), Ch B (Scale = 5 V/Div, Ypos = 0, AC), Trigger (Pos edge, Level = 0, Nor, A). Click the On-Off switch and run the simulation to completion (Tran = 0.210 s). Record the total ac rms current (I), the ac rms voltage across the inductor (V_L), and the total ac rms voltage across the R-L network (V).

I = __________ rms V_L = __________ rms V = __________ rms

Step 8. Based on the readings in Step 7, calculate the real ac power (P) in the R-L circuit.

Step 9. Based on the readings in Step 7, calculate the reactive power (Q) in the R-L circuit.

Step 10. Based on the readings in Step 7, calculate the apparent power (S) in the R-L circuit.

Step 11. Based on the calculated real power (P), the reactive power (Q), and the apparent power (S) in Steps 8–10, draw the power triangle and determine the power factor (pf) for the R-L network.

Step 12. The oscilloscope is plotting voltage V (red curve plot) and current I (blue curve plot). Using the oscilloscope curve plots, determine the phase (θ) between voltage V and current I. Based on the value of the phase (θ) calculate the power factor (pf).

Question: How did your measured power factor in Step 12 compare with the value determined from the power triangle in Step 11? Is it lagging or leading? **Explain why**.

Step 13. Pull down the File menu and open FIG23-3. Bring down the oscilloscope enlargement and make sure that the following settings are selected: Time base (Scale = 2 ms/Div, Xpos = 0, Y/T), Ch A (Scale = 10 V/Div, Ypos = 0, AC), Ch B (Scale = 5 V/Div, Ypos = 0, AC), Trigger (Pos edge, Level = 0, Nor, A). Click the On-Off switch and run the simulation to completion (Tran = 0.21 s). Record the total ac rms current (I), the total ac rms voltage across the R-L-C network (V), the ac rms capacitor voltage (V_C), and the ac rms inductor voltage (V_L).

I = __________ rms V = __________ rms

V_C = __________ rms V_L = __________ rms

Step 14. Based on the readings in Step 13, calculate the real ac power (P) in the R-L-C circuit.

Step 15. Based on the readings in Step 13, calculate the reactive power (Q) in the R-L-C circuit.

Step 16. Based on the readings in Step 13, calculate the apparent power (S) in the R-L-C circuit.

Step 17. Based on the calculated real power (P), the reactive power (Q), and the apparent power (S) in Steps 14–16, draw the power triangle and determine the power factor (pf) for the R-L-C network.

Step 18. The oscilloscope is plotting voltage V (red curve plot) and current I (blue curve plot). Using the oscilloscope curve plots, determine the phase (θ) between voltage V and current I. Based on the value of the phase (θ), calculate the power factor (pf).

Question: How did your measured power factor in Step 18 compare with the value determined from the power triangle in Step 17? Is it lagging or leading? **Explain why**.

Step 19. Pull down the File menu and open FIG23-4. Bring down the oscilloscope enlargement and make sure that the following settings are selected: Time base (Scale = 2 ms/Div, Xpos = 0, Y/T), Ch A (Scale = 10 V/Div, Ypos = 0, AC), Ch B (Scale = 10 mV/Div, Ypos = 0, AC), Trigger (Pos edge, Level = 0, Nor, A). In this part of the experiment you will calculate the value of C needed to make the power factor close to unity. You will then change C to the calculated value, connect it to the circuit, and run the simulation again to determine if the correct value for C was calculated. Click the On-Off switch and run the simulation to completion (Tran = 0.21 s) without C connected. Record the total ac rms current entering the R-L circuit (I), current I_{AB}, and voltage V_{AB}.

$I =$ __________ rms $\quad I_{AB} =$ __________ rms $\quad V_{AB} =$ __________ rms

Step 20. The oscilloscope is plotting voltage V_{AB} (red curve plot) and current I (blue curve plot). Using the oscilloscope curve plots, determine the phase (θ) between voltage V_{AB} and current I.

Step 21. Based on the readings in Step 19, calculate the real ac power (P) in the R-L circuit.

Step 22. Based on the readings in Step 19, calculate the apparent power (S) in the R-L circuit.

Step 23. Based on the real power (P) and the apparent power (S), calculate the power factor (pf) and the power factor angle (θ) for the R-L network.

Question: How did the power factor angle (θ) in Step 23 compare with the phase between V_{AB} and I in Step 20? Is this expected? **Explain.**

Step 24. Based on the power factor angle (θ), the real power (P), and the apparent power (S), draw the power triangle and determine the reactive power (Q).

Step 25. Based on the reactive power (Q) and the voltage V_{AB}, calculate the capacitive reactance (X_C) needed for power factor correction.

Step 26. Based on the reactance calculated in Step 25 and the frequency, calculate the capacitance (C) needed for power factor correction (to make the power factor close to unity).

Step 27. Change the value of capacitor C to the value calculated in Step 26 and connect it between terminals A-B.

(*Note:* If you are performing this experiment in a hardwired laboratory, use a capacitor as close as possible to the calculated value.)

Step 28. Click the On-Off switch to run the simulation to completion again. Record the total ac rms current entering the circuit (I), current I_{AB}, and voltage V_{AB}.

I = __________ rms I_{AB} = __________ rms V_{AB} = __________ rms

Step 29. Based on the readings in Step 28, calculate the real ac power (P) dissipated in the R-L circuit.

Question: How did the real power calculated in Step 29 compare with the real power calculated in Step 21? **Explain**.

Step 30. Based on the value of I and V_{AB} in Step 28, calculate the apparent power (S).

Step 31. Based on the real power (P) calculated in Step 29 and the apparent power (S) calculated in Step 30, calculate the new power factor (pf).

Questions: How does the new power factor compare with the power factor calculated in Step 23? Is it close to unity? What is the difference between the real power and the apparent power? Was power factor correction accomplished?

Based on the oscilloscope readings, are the circuit current I and the voltage V_{AB} in-phase? Is it what you expected? **Explain.**

Troubleshooting Problem

1. Pull down the File menu and open FIG23-5. Click the On-Off switch and run the simulation to completion (Tran = 0.339 s). Determine the capacitance C needed for power factor correction. Next, change the value of C in the circuit window to your calculated value and connect it between terminals A-B. Click the On-Off switch, run the simulation to completion, and determine if the capacitance is the correct value for power factor correction (V and I should be in phase on the oscilloscope, making angle θ equal to zero degrees and the power factor equal to unity).

   Hint: First, determine the power factor angle (θ) for the motor by measuring the phase difference between the motor voltage V (red curve plot) and motor current I (blue curve plot) on the oscilloscope screen. Determine the apparent power (S) from the rms motor current (I) and the rms motor voltage (V). Determine the reactive power (Q) from the power triangle and determine the value of C from the reactive power.

EXPERIMENT

Name______________________________

Date______________________________

Nodal Analysis of AC Circuits

Objectives:

1. Solve for the nodal voltage in a two-node ac circuit and compare your calculated value with the measured value.
2. Solve for the nodal voltages in a three-node ac circuit and compare your calculated values with the measured values.

Materials:

One ac signal generator
Two 0–10 V ac voltmeters
Four 0.2 μF capacitors
One 100 mH inductor
Resistors—600 Ω, 775 Ω, and 1 kΩ

Theory:

The **nodal voltage method** of analyzing ac circuits requires using **Kirchhoff's current law** to determine the voltage at each circuit node (junction) with respect to a previously selected reference node, called the ground node. The branch impedances are represented by **complex numbers**, each with a real part and an imaginary part. The real part represents the branch resistance and the imaginary part represents the branch reactance ($+jX_L$ for inductive reactance and $-jX_C$ for capacitive reactance). The steps required to solve for the nodal voltages are as follows:

1. Arbitrarily assign a reference node to be indicated with the ground symbol, normally at the bottom of the circuit.

2. Convert each voltage source and impedance to its equivalent current source and impedance. This can be done only if there is an impedance in series with the source. (This step is not required, but it will make the calculations easier.)

3. Arbitrarily assign unknown voltage symbols ($V_1, V_2, \ldots, V_n$) to each of the nodes, except for the reference node. The calculated nodal voltage values will be with respect to the reference node. If a voltage source is connected directly between two nodes without any impedance in the branch, assign a voltage symbol to one of the nodes and represent the other nodal voltage in terms of the first nodal voltage and the source voltage. If a voltage source is connected directly

between a node and the reference node without any impedance in the branch, then the nodal voltage is equal to the source voltage.

4. Arbitrarily assign current directions to each circuit branch in which there is no current source. Normally the current directions are toward the reference node.

5. Apply Kirchhoff's current law to each node (except the reference node) and represent each branch current in terms of the unknown nodal voltages and branch impedances. You will obtain one complex equation at each node (except the reference node). Therefore, if there are n nodes, you will obtain n – 1 complex equations. If there is a voltage source between two nodes without any resistance in the branch, treat the two nodes as one supernode.

6. Solve the resulting complex equations simultaneously to obtain the nodal voltage values.

Before completing the experiment, you may need to review the nodal method of circuit analysis and **phasors** represented as complex numbers in your classroom textbook. It would also be helpful to review **source transformation** (converting equivalent voltage sources to equivalent current sources).

In this experiment you will analyze the two-node circuit in Figure 24-1 and the three-node circuit in Figure 24-2.

> ***Note:* This experiment is for advanced students and may be skipped without a loss in continuity.**

> *Note:* If you do the experiment in a real laboratory environment (hardwired), see the end of the Theory section of Experiment 19 regarding the inductor resistance (R_L).

Figure 24-1 Nodal Analysis—Two-Node AC Circuit

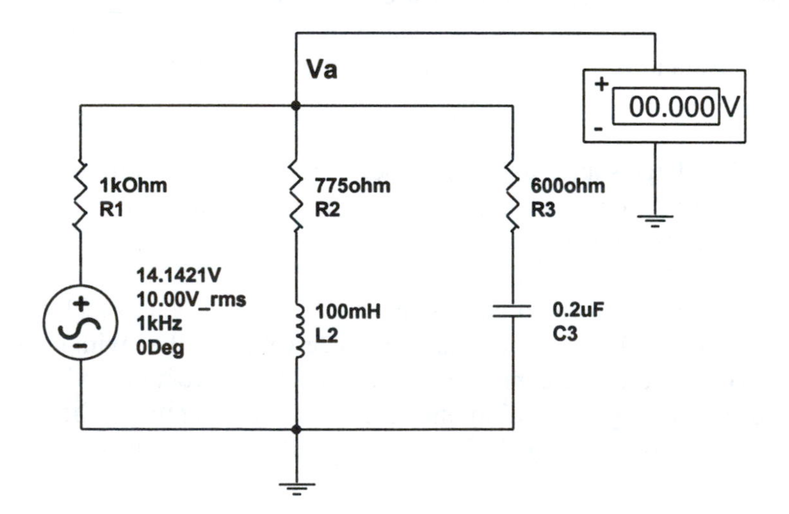

Figure 24-2 Nodal Analysis—Three-Node AC Circuit

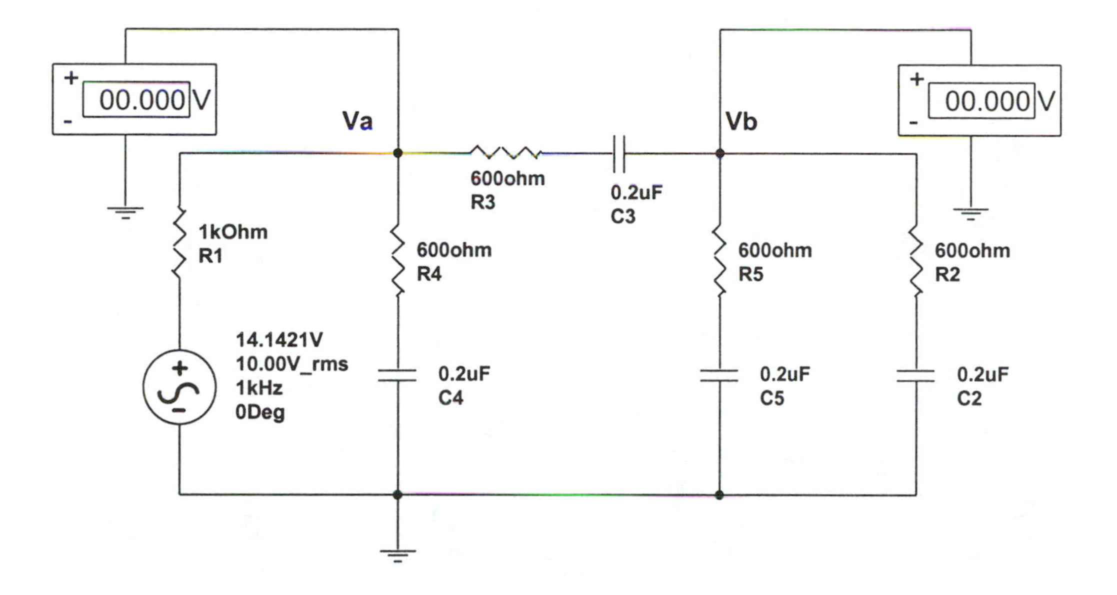

Procedure:

Step 1. Pull down the File menu and open FIG24-1. Click the On-Off switch to run the simulation to completion. Record ac rms nodal voltage V_a.

V_a = __________

Step 2. Draw the phasor circuit and solve for the nodal voltage (V_a) for the circuit in Figure 24-1. Use the space below and on the next page.

Question: How did your calculated value for V_a compare with the measured value in Step 1?

Step 3. Pull down the File menu and open FIG24-2. Click the On-Off switch to run the simulation to completion. Record ac rms nodal voltages V_a and V_b.

V_a = __________ V_b = __________

Step 4. Draw the phasor circuit and solve for nodal voltages V_a and V_b for the circuit in Figure 24-2.

Question: How did your calculated values for V_a and V_b compare with the measured values in Step 3?

Name________________________________

Date________________________________

EXPERIMENT

25

Mesh Analysis of AC Circuits

Objectives:

1. Solve for the ac rms mesh currents in a two-loop ac circuit.
2. Based on the mesh current values, determine the ac rms current in each branch and compare your calculated values with the measured values.

Materials:

One ac signal generator
Three 0–20 mA ac milliammeters
One 0.2 μF capacitor
One 95 mH inductor
One 600 Ω resistor

Theory:

The **mesh current method** of analyzing ac circuits requires using **Kirchhoff's voltage law** to sum ac voltages around a closed path in terms of the mesh (loop) current variables. The branch impedances are represented by **complex numbers**, each with a real part and an imaginary part. The real part represents the branch resistance and the imaginary part represents the branch reactance ($+jX_L$ for inductive reactance and $-jX_C$ for capacitive reactance). The steps required to solve for the ac mesh current phasors are as follows:

1. Arbitrarily assign a clockwise ac current phasor to each interior closed loop (mesh) and label it with an unknown variable ($I_1, I_2, \ldots, I_n$). If a current source is in a loop that is not in common with another loop, then the mesh current is equal to the source current, taking direction into account. If a current source is in a loop that is in common with another loop, the sum of the mesh currents is equal to the source current, taking direction into account.

2. Assign voltage polarities across all of the impedance elements in the circuit based on the mesh current directions. For an impedance element that is in common with two loops, the polarity of the voltage due to each mesh current should be indicated on the appropriate side of the impedance element.

3. Apply Kirchhoff's voltage law to each loop to develop the ac mesh current equations. You will obtain one equation for each loop, except for those loops where there is a current source. If there are current sources in the circuit, you will obtain one equation from each source.

4. Solve the resulting equations simultaneously to obtain the ac rms mesh current phasor values.

5. A branch that is in common with more than one loop has a branch current that is determined by algebraically combining the ac mesh current phasors that are common to the branch.

Before completing this experiment, you may need to review the mesh current method of circuit analysis and **phasors** represented as **complex numbers** in your classroom textbook.

In this experiment you will analyze the two-loop circuit in Figure 25-1 for the ac rms mesh current values and then calculate the branch current values from the mesh current values.

> ***Note:*** **This experiment is for advanced students and may be skipped without a loss in continuity.**

> *Note:* If you do the experiment in a real laboratory environment (hardwired), see the end of the Theory section of Experiment 19 regarding the inductor resistance (R_L).

Figure 25-1 Mesh Current Analysis—Two-Loop AC Circuit

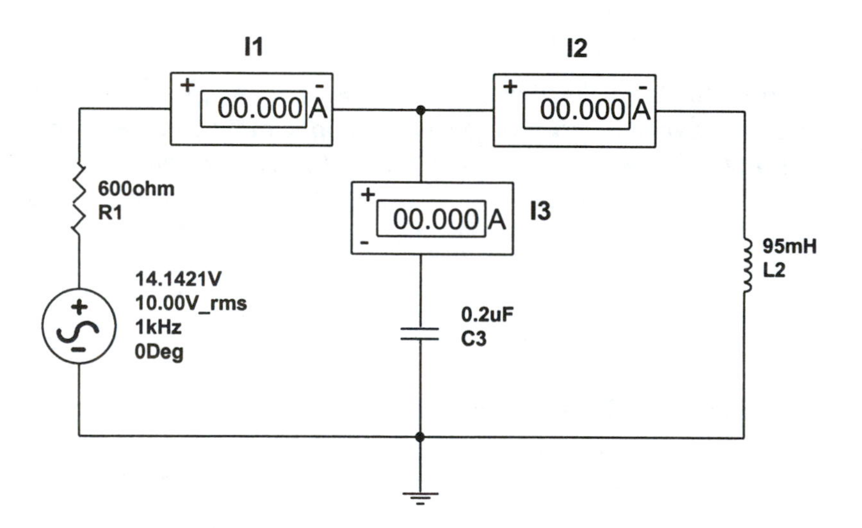

Procedure:

Step 1. Pull down the File menu and open FIG25-1. Click the On-Off switch to run the simulation to completion. Record the ac rms branch currents I_1, I_2, and I_3.

I_1 = __________ I_2 = __________ I_3 = __________

Step 2. Draw the phasor circuit and solve for the ac rms mesh (loop) currents for the circuit in Figure 25-1.

Step 3. From the values of the ac rms mesh currents calculated in Step 2, calculate the values of the ac rms branch currents I_1, I_2, and I_3.

Question: How did your calculated values for the ac rms branch currents I_1, I_2, and I_3 compare with the measured values in Step 1?

EXPERIMENT

26

Name_______________________________

Date_______________________________

Thevenin Equivalent for AC Circuits

Objectives:

1. Determine the ac Thevenin equivalent circuit for a known ac circuit.
2. Prove the validity of Thevenin's theorem for ac circuits.

Materials:

One ac signal generator
One dual-trace oscilloscope
One 0–10 V ac voltmeter
One 0–20 mA ac milliammeter
One 0.2 μF capacitor
One 100 mH inductor
Resistors—1 Ω, 500 Ω, 1 kΩ

Theory:

Thevenin's Theorem

Any linear two-terminal ac network of fixed impedances and ac voltage sources may be replaced with a single ac voltage source in series with a single impedance. The single ac voltage source is called the **Thevenin voltage (V_{TH})** and its magnitude is equal to the **ac open circuit voltage (V_{oc})** at the terminals of the original network. The ac open circuit voltage (V_{oc}) for the circuit in Figure 26-1 can be calculated by first determining the ac circuit current (I). The ac rms circuit current (I) is calculated by dividing the ac rms source voltage (V) by the total impedance (Z_T).

Therefore,

$$I = \frac{V}{Z_T}$$

where

$$Z_T = \sqrt{(R_1 + R_2)^2 + X_{L1}^2}$$

and

$$X_{L1} = 2\pi f L_1$$

The ac open circuit voltage (at terminals a-b) can be calculated from the ac circuit current (I) as follows:

$$V_{oc} = IR_2$$

The single impedance is called the **Thevenin impedance (Z_{TH})** and has a magnitude equal to the ac open circuit voltage (V_{oc}) at terminals a-b divided by the **ac short circuit current (I_{sc})** between terminals a-b of the original network. Therefore,

$$Z_{TH} = \frac{V_{oc}}{I_{sc}}$$

The ac short circuit current magnitude (I_{sc}) is measured by connecting an ac ammeter between terminals a-b of the original network and recording the current reading, as shown in Figure 26-2. (*Note:* An ammeter has a very low internal resistance and can be considered to be a short.) The 1 Ω resistor makes it possible to measure I_{sc} using the oscilloscope and can be considered to be a short circuit. The ac short circuit current (I_{sc}) in Figure 26-2 is calculated by drawing a short between terminals a-b of the original network and calculating the ac current in the short. Because the short has placed the ac source voltage (V) across impedance Z_1 (R_1 and L_1), and resistor R_2 is shorted, the short circuit current (I_{sc}) can be calculated from:

$$I_{sc} = \frac{V}{Z_1}$$

where

$$Z_1 = \sqrt{R_1^2 + X_{L1}^2}$$

The ac Thevinen impedance (Z_{TH}) can also be determined by removing all voltage and current sources from the original network (voltage sources are replaced with a short and current sources are replaced with an open) and determining the resulting impedance at terminals a-b of the network.

By converting the Thevenin impedance (Z_{TH}) from **polar form** ($Z_{TH}\angle \pm\theta$) to **rectangular form** ($R_{TH} \pm jX_{TH}$), the resistance (R_{TH}) and the reactance (X_{TH}) in the Thevenin equivalent can be determined.

When an impedance Z_L (R_L and C_L) is connected between terminals a-b of the circuit, as shown in Figure 26-3, the ac voltage across the terminals (V_{ab}) of the original circuit will be the same as the ac voltage across the terminals of the Thevenin equivalent circuit if the same impedance (Z_L) is connected across the Thevenin equivalent. In Figure 26-3, the phase (α) of the load voltage (V_{ab}) with respect to the load current (I_L) can be found from the load capacitive reactance (X_{CL}) and the load resistance (R_L) as follows:

$$\alpha = -\arctan\frac{X_{CL}}{R_L}$$

Review the polar and rectangular representation of phasors and impedance in your text before proceeding with this experiment. Also review how to determine the phase difference between periodic functions by measuring the time difference between them in the Theory section of Experiment 21.

> ***Note:*** **This experiment is for advanced students and may be skipped without a loss in continuity.**

> *Note:* If you do the experiment in a real laboratory environment (hardwired), see the end of the Theory section of Experiment 19 regarding the inductor resistance (R_L).

Figure 26-1 Determining the AC Open Circuit Voltage

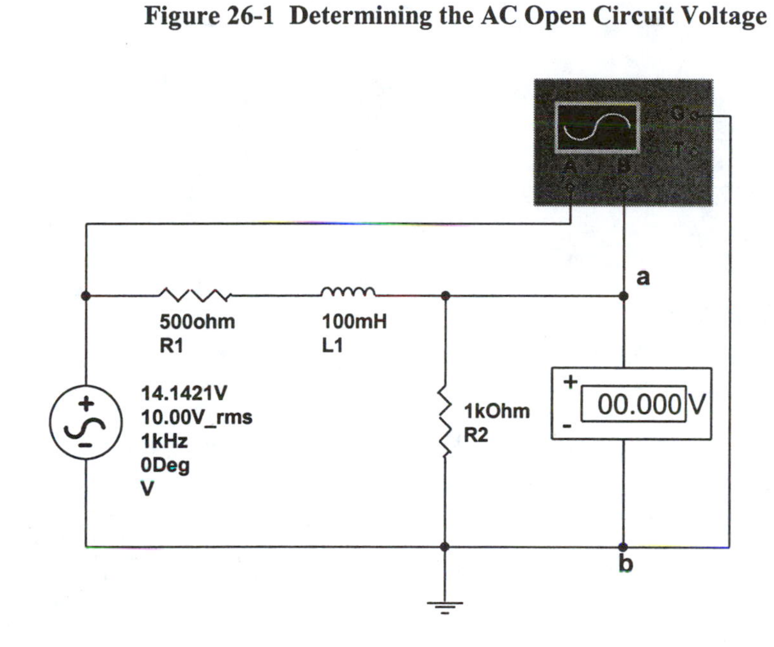

Procedure:

Step 1. Pull down the File menu and open FIG26-1. Bring down the oscilloscope enlargement and make sure that the following settings are selected: Time base (Scale = 200 μs/Div, Xpos = 0, Y/T), Ch A (Scale = 5 V/Div, Ypos = 0, AC), Ch B (Scale = 5 V/Div, Ypos = 0, AC), Trigger (Pos edge, Level = 0, Sing, A). Click the On-Off switch to run the simulation. Record the ac rms open circuit voltage (V_{oc}) across terminals a-b.

V_{oc} = __________

Figure 26-2 Determining the AC Short Circuit Current

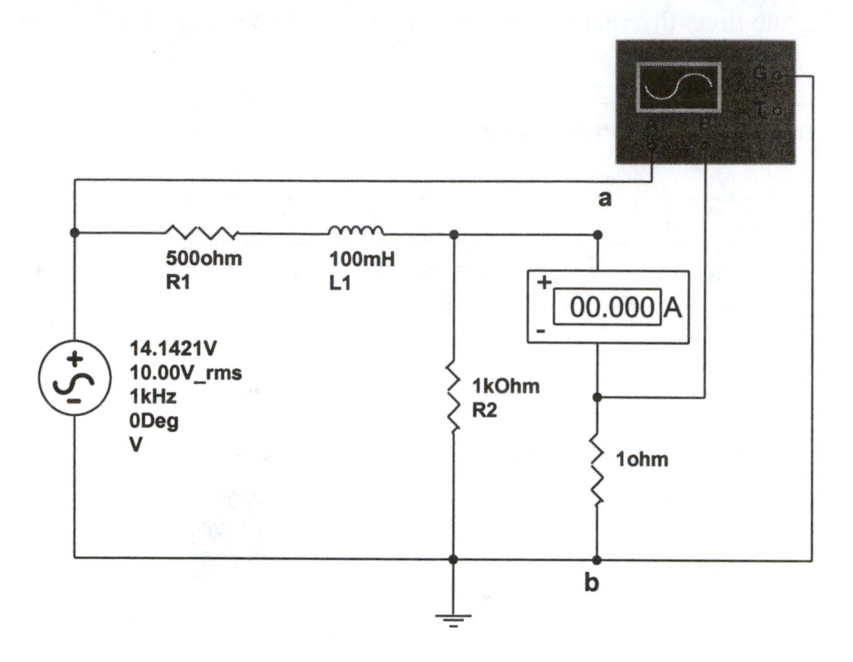

Figure 26-3 Determining AC Voltage V_{ab} with Z_L Added

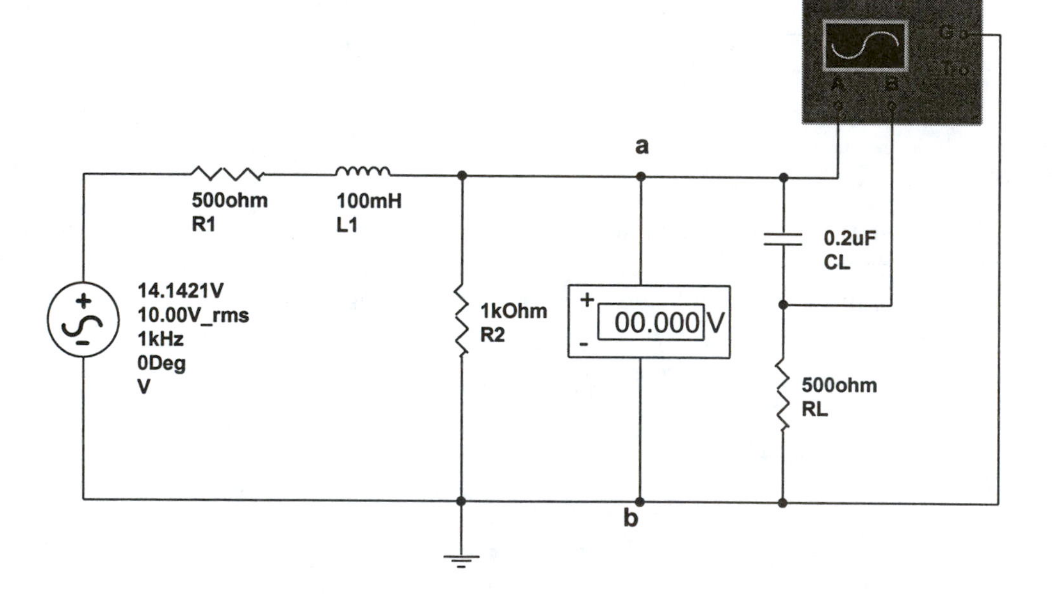

Step 2. Based on the oscilloscope readings, determine the phase (θ) of V_{oc} (blue curve plot) using V (red curve plot) as the reference at zero degrees.

θ = __________

Step 3. Based on the circuit component values in Figure 26-1, calculate the ac rms open circuit voltage magnitude (V_{oc}) across terminals a-b.

Question: How did your calculated value of the magnitude of V_{oc} compare with the magnitude of V_{oc} measured in Step 1?

Step 4. Pull down the File menu and open FIG26-2. Bring down the oscilloscope enlargement and make sure that the following settings are selected: Time base (Scale = 200 μs/Div, Xpos = 0, Y/T), Ch A (Scale = 5 V/Div, Ypos = 0, AC), Ch B (Scale = 10 mV/Div, Ypos = 0, AC), Trigger (Pos edge, Level = 0, Sing, A). Click the On-Off switch to run the simulation. Record the ac rms short circuit current (I_{sc}) between terminals a-b. (The 1 Ω resistor is practically a short circuit.)

I_{sc} = __________

Step 5. Based on the oscilloscope readings, determine the phase (θ) of I_{sc} (blue curve plot) using V (red curve plot) as the reference at zero degrees.

θ = __________

Step 6. Based on the circuit component values in Figure 26-2, calculate the ac rms short circuit current magnitude (I_{sc}) between terminals a-b.

Question: How did your calculated value of the magnitude of I_{sc} compare with the magnitude of I_{sc} measured in Step 4?

Step 7. Based on the magnitude of V_{oc}, determine the magnitude of the Thevenin voltage (V_{TH}).

Step 8. Based on the magnitude and phase of V_{oc} and I_{sc}, determine the Thevenin impedance (Z_{TH}) in polar and rectangular form.

Step 9. Based on the values calculated in Steps 7 and 8, draw the ac Thevenin equivalent circuit. Represent the impedance as a resistance and a reactance.

Step 10. Pull down the File menu and open FIG26-3. Bring down the oscilloscope enlargement and make sure that the following settings are selected: Time base (Scale = 200 μs/Div, Xpos = 0, Y/T), Ch A (Scale = 5 V/Div, Ypos = 0, AC), Ch B (Scale = 2 V/Div, Ypos = 0, AC), Trigger (Pos edge, Level = 0, Sing, A). Click the On-Off switch to run the simulation. Record the ac rms voltage V_{ab}.

V_{ab} = __________

Step 11. Based on the oscilloscope readings, determine the phase (α) of V_{ab} (red curve) with respect to I_L (blue curve).

α = __________

Step 12. Draw the circuit in Figure 26-3 with the section to the left of terminals a-b replaced with the ac Thevenin equivalent circuit in Step 9. Use this circuit to solve for the magnitude of the ac voltage V_{ab}.

Question: How did your magnitude of V_{ab} in Step 12 compare with the measured value for V_{ab} in Step 10? Was the ac Thevenin equivalent circuit equivalent to the original ac circuit? **Explain.**

Step 13. Based on the load resistance (R_L) and load capacitive reactance (X_{CL}), calculate the phase (α) of V_{ab} with respect to I_L.

Question: How did your calculated value of the phase in Step 13 compare with the phase determined in Step 11?

Name______________________________

Experiment

Date______________________________

27

Series Resonance

Objectives:

1. Determine the resonant frequency of a series resonant circuit and compare the measured value with the calculated value.
2. Determine the bandwidth of a series resonant circuit and compare the measured value with the calculated value.
3. Determine the quality factor for a series resonant circuit.
4. Determine the impedance of a series resonant circuit at the resonant frequency.
5. Determine the phase relationship between the voltage and current of a series resonant circuit at the resonant frequency.
6. Determine the effect of changing the circuit resistance on the resonant frequency and the bandwidth of a series resonant circuit.

Materials:

One function generator
One dual-trace oscilloscope
One 100 mH inductor
One 0.25 μF capacitor
One 1 kΩ resistor

Theory:

Complete Experiment 21 before attempting this experiment. If that experiment has been completed, review the Theory section.

Frequency selectivity, which makes it possible to select a frequency transmitted by a particular radio or TV station and reject other frequencies, is based on the principal of **resonance**. In a series R-L-C circuit, such as shown in Figures 27-1 and 27-2, the capacitive reactance varies inversely with frequency ($X_C = 1/2\pi fC$) and the inductive reactance varies directly with frequency ($X_L = 2\pi fL$). As the frequency of a sinusoidal signal applied to a series R-L-C circuit increases from zero, the capacitive reactance decreases and the inductive reactance increases. At a particular frequency, the capacitive and inductive reactances will be equal. This frequency is called the **resonant frequency (f_r)**. Because the inductive reactance and the capacitive reactance phasors are 180 degrees out-of-phase, the total reactance is zero at the resonant frequency ($X = X_L - X_C = 0$). Therefore, at the resonant frequency of a series resonant circuit the total **impedance** is at its lowest value and is equal to the circuit resistance (R). This causes the circuit current (I) to be at its maximum value. Therefore, at resonance

$$X_L = X_C$$

$$Z = \sqrt{R^2 + (X_L - X_C)^2} = \sqrt{R^2} = R$$

$$2\pi f_r L = \frac{1}{2\pi f_r C}$$

Solving for f_r produces the equation

$$f_r = \frac{1}{2\pi\sqrt{LC}}$$

for determining the resonant frequency.

As the sinusoidal signal frequency is reduced below the resonant frequency (f_r), the impedance (Z) will increase because the inductive reactance will decrease and the capacitive reactance will increase, causing the total reactance ($X = X_L - X_C$) to increase in magnitude. This will cause the circuit current (I) to decrease. As the sinusoidal signal frequency is increased above the resonant frequency, the impedance will increase again because the inductive reactance will increase and the capacitive reactance will decrease, causing the total reactance to increase again. This will cause the circuit current to decrease again. This means that, in a series resonant circuit, the impedance will be at its minimum value and the circuit current will be at its maximum value at the resonant frequency.

Because the impedance is resistive at the resonant frequency ($X = X_L - X_C = 0$), the circuit voltage and current are in phase ($\theta = 0°$). As the frequency is decreased below resonance, the capacitive reactance will increase and the inductive reactance will decrease, making the circuit capacitive and causing the phase (θ) to decrease towards –90 degrees (V lags I). As the frequency is increased above resonance, the capacitive reactance will decrease and the inductive reactance will increase, making the circuit inductive and causing the phase to increase towards +90 degrees (V leads I).

The **bandwidth (BW)** of a series resonant circuit can be measured from the frequency curve plot of the current (I) by determining the low frequency (f_l) and the high frequency (f_h) at which the current (I) drops to 0.707 (–3dB) of the peak value. The bandwidth can be calculated from the equation

$$BW = f_h - f_l$$

The **quality factor (Q)** measures how narrow the resonant circuit bandwidth (BW) is in relation to the resonant frequency (f_r). The higher the quality factor, the more narrow the bandwidth in relation to the resonant frequency. The quality factor (Q) can be calculated from

$$Q = \frac{f_r}{BW}$$

The quality factor (Q) is also a measure of the ratio of the energy stored in the inductor or capacitor to the energy dissipated in the circuit resistance. This can also be expressed as the ratio of the reactive power to the real power. Therefore,

$$Q = \frac{I^2X_L}{I^2R} = \frac{X_L}{R}$$

where X_L is the inductive reactance at the resonant frequency (f_r).

The bandwidth (BW) can be calculated from the circuit component values using the equation

$$BW = \frac{f_r}{Q} = \frac{f_r}{X_L/R} = \frac{f_rR}{2\pi f_rL} = \frac{R}{2\pi L}$$

Note: If you do the experiment in a real laboratory environment (hardwired), see the end of the Theory section of Experiment 19 regarding the inductor resistance (R_L).

Figure 27-1 Series Resonance

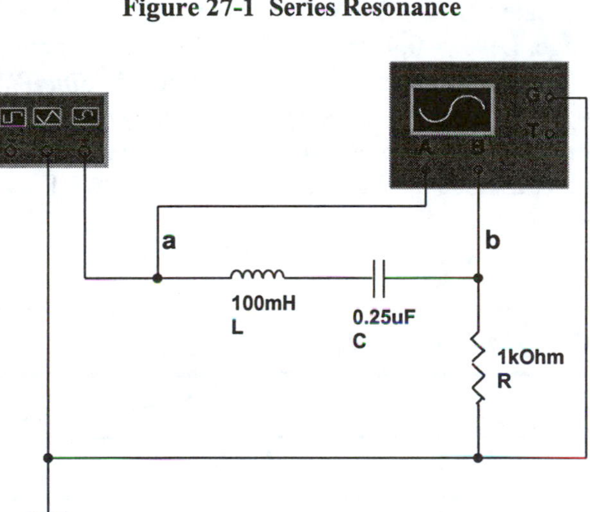

Figure 27-2 Series Resonance Frequency Plot

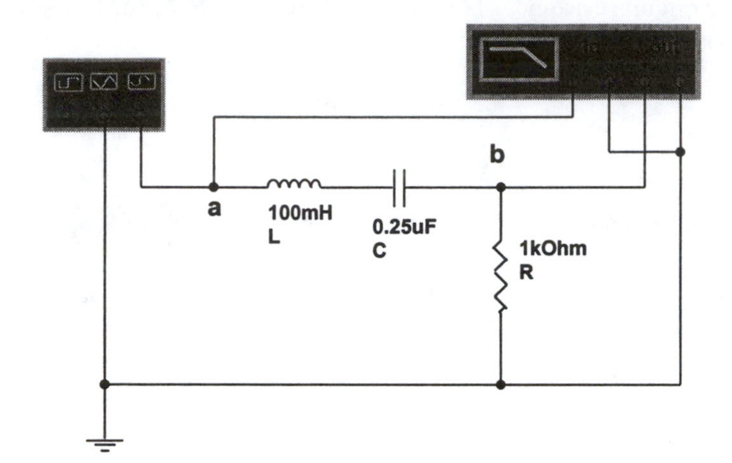

Procedure:

Step 1. Pull down the File menu and open FIG27-1. Bring down the oscilloscope enlargement and make sure that the following settings are selected: Time base (Scale = 200 μs/Div, Xpos = 0, Y/T), Ch A (Scale = 10 V/Div, Ypos = 0, AC), Ch B (Scale = 5 V/Div, Ypos = 0, AC), Trigger (Pos edge, Level = 0, Nor, A). Bring down the function generator enlargement and make sure that the following settings are selected: *Sine Wave*, Freq = 1 kHz, Ampl = 10 V, Offset = 0. Click the On-Off switch to run the simulation. Click "Pause" when steady state is reached on the oscilloscope. Record the peak voltage V_a at node a (red curve plot) and V_b at node b (blue curve plot) in Table 27-1 at the appropriate frequency.

Table 27-1

f (Hz)	V_a (V)	V_b (V)	I (mA)	Z (kΩ)
100				
300				
700				
1000				
3000				
4000				
10000				

Step 2. Change the frequency of the function generator to each frequency in Table 27-1, run the simulation, and record each peak value of V_a and V_b for each frequency. Adjust the oscilloscope as needed.

Step 3. Based on each value of V_b in Table 27-1 and R in Figure 27-1, calculate the current (I) at each frequency and record your answers in the table.

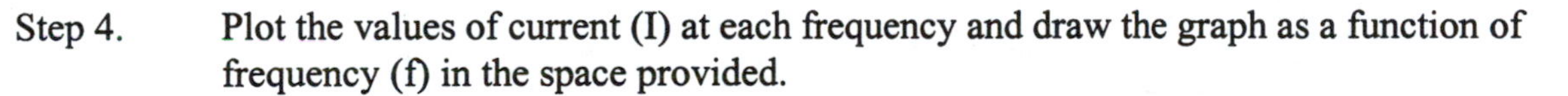

Step 4. Plot the values of current (I) at each frequency and draw the graph as a function of frequency (f) in the space provided.

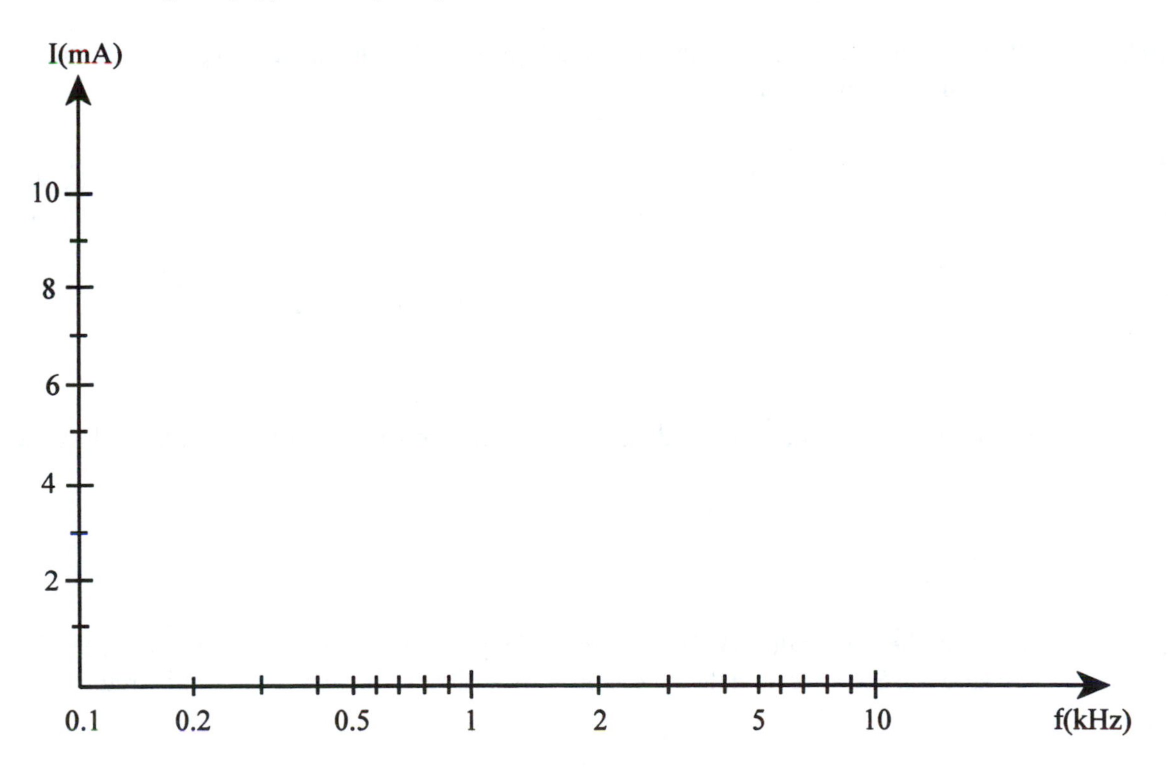

Step 5. Based on the curve plot in Step 4, determine the resonant frequency (f_r) for the series resonant circuit.

f_r = __________

Step 6. Based on the circuit component values in Figure 27-1, calculate the expected resonant frequency (f_r) of the series resonant circuit.

Question: How did your calculated resonant frequency in Step 6 compare with the resonant frequency determined from the curve?

Step 7. Based on the curve plot in Step 4, determine the bandwidth (BW) of the series resonant circuit.

BW = __________

Step 8. Based on the circuit component values in Figure 27-1, calculate the expected bandwidth (BW) of the series resonant circuit.

Question: How did your calculated bandwidth in Step 8 compare with the bandwidth from the curve in Step 7?

Step 9. Based on the bandwidth (BW) measured in Step 7 and the resonant frequency (f_r) measured in Step 5, calculate the quality factor (Q) for this series resonant circuit.

Step 10. Based on each value of V_a and I in Table 27-1, calculate the impedance (Z) of the series resonant circuit at each frequency and record your answers in the table.

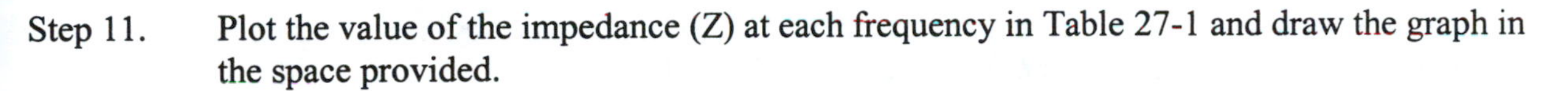

Step 11. Plot the value of the impedance (Z) at each frequency in Table 27-1 and draw the graph in the space provided.

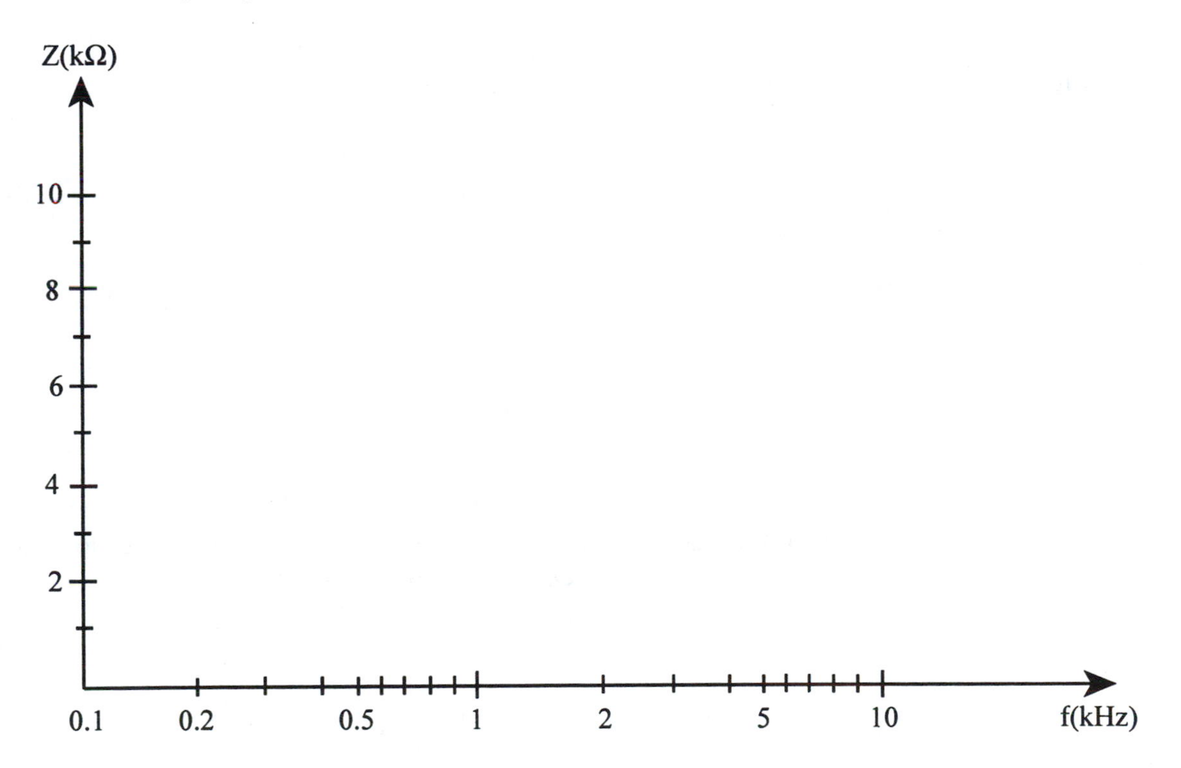

Question: Based on the graph in Step 11, what conclusion can you draw about the impedance of the series resonant circuit as the frequency varies? **Explain.**

Step 12. Based on the circuit component values in Figure 27-1, calculate the impedance of the series resonant circuit at the resonant frequency.

Questions: How did your calculated impedance at the resonant frequency in Step 12 compare with the impedance at the resonant frequency recorded in Table 27-1?

What was the relationship between the impedance at the resonant frequency and the value of R? **Explain.**

Step 13. Change the frequency of the function generator to the resonant frequency (f_r), run the simulation, and record the phase difference (θ) between the voltage (red curve plot) and current (blue curve plot). Change the oscilloscope settings as needed.

θ = __________

Question: What conclusion can you draw about the phase difference between the voltage and the current at resonance? **Explain.**

> *Note:* Steps 14–16 should be performed on the computer using Multisim because a Bode plotter may not be available in a hardwired laboratory.

Step 14. Pull down the File menu and open FIG27-2. Bring down the Bode plotter enlargement and make sure that the following settings are selected: *Magnitude*, Vertical (Lin, F = 1, I = 0), Horizontal (Log, F = 100 kHz, I = 10 Hz). Click the On-Off switch to run the simulation until the Bode plot is complete, then "pause" the simulation. The Bode plotter is plotting the series resonant circuit current (I) as a function of frequency. Unity (1) represents a current (I) of 10 mA. Measure and record the resonant frequency (f_r) and the bandwidth (BW) of the curve plot.

f_r = __________ BW = __________

Step 15. Click "Phase" on the Bode plotter. On the Bode plotter vertical axis, set F to 90° and set I to –90°. Move the cursor to the resonant frequency (f_r) and record the phase (θ).

θ = ___________

Questions: Was the phase (θ) between the input voltage and the circuit current what you expected? What did the phase do above and below the resonant frequency?

Step 16. Change the resistance (R) to 100 Ω and repeat Step 14. Change the Bode plotter settings as needed.

f_r = __________ BW = __________

Question: What conclusion can you draw about the effect of the resistance value on the resonant frequency and the bandwidth?.

Troubleshooting Problems

1. Pull down the File menu and open FIG27-3. Bring down the Bode plotter enlargement, click the On-Off switch to run the simulation, then "pause" the simulation after the Bode plot is plotted. Based on the frequency plot on the Bode plotter, determine the value of R and C.

   R = __________ C = __________

2. Pull down the File menu and open FIG27-4. Bring down the Bode plotter enlargement, click the On-Off switch to run the simulation, then "pause" the simulation after the Bode plot is plotted. Based on the frequency plot on the Bode plotter, determine the value of L and C.

   L = __________ C = __________

Name______________________________

Date______________________________

Parallel Resonance

Objectives:

1. Determine the resonant frequency of a parallel resonant circuit and compare the measured value with the calculated value.
2. Determine the bandwidth of a parallel resonant circuit and compare the measured value with the calculated value.
3. Determine the quality factor of a parallel resonant circuit.
4. Determine the impedance of a parallel resonant circuit at the resonant frequency and compare the measured value with the calculated value.
5. Determine the phase relationship between the voltage and current of a parallel resonant circuit at the resonant frequency.
6. Determine the effect of resistance on the resonant frequency and the bandwidth of a parallel resonant circuit.

Materials:

One function generator
One dual-trace oscilloscope
One 100 mH inductor
One 0.25 μF capacitor
Resistors—10 Ω and 20 kΩ

Theory:

Complete Experiments 22 and 27 before attempting this experiment. If those experiments have been completed, review the Theory sections.

A **parallel resonant circuit** consists of a capacitor in parallel with an inductor and is often referred to as a **tank circuit**. Due to the high **quality factor (Q)** and frequency response, tank circuits are used extensively in communications equipment, such as radio and TV receivers and transmitters. In the parallel resonant circuits in Figures 28-1 and 28-2, resistor R_W has been added to simulate the coil wire resistance of inductor L. **If you are performing this experiment in a hardwired lab environment, use the actual value of the inductor coil resistance for the value of R_W.** Resistor R has been added to simulate a load resistance across the tank circuit. The 10 Ω resistor is not part of the tank circuit, but is for the purpose of measuring the current entering and leaving the tank circuit. The voltage across the 10 Ω resistor (V_b) is proportional to the current. Therefore, Channel B on the oscilloscope is plotting the current (I) entering and leaving the tank circuit ($I = V_b/10$) and Channel A is plotting the voltage across the tank circuit (V_a).

Because the inductor **wire resistance (R_W)** is in series with inductor L, the circuit is not exactly a parallel resonant circuit. In order to make the circuit a true parallel resonant circuit, the series combination of inductance L and resistance R_W must first be converted into an equivalent parallel network with a resistance (R_{eq}) in parallel with an inductance (L_{eq}). In Figures 28-1 and 28-2, the **parallel equivalent resistance (R_{eq})** will be in parallel with resistor R, making the total resistance of the parallel resonant circuit (R_P) equal to the parallel equivalent of resistors R and R_{eq}. Therefore,

$$R_P = \frac{(R)(R_{eq})}{R + R_{eq}}$$

The equation for converting resistance R_W to equivalent parallel resistance R_{eq} is

$$R_{eq} = R_W(Q_L{}^2 + 1)$$

and the equation for converting inductance L to equivalent parallel inductance L_{eq} is

$$L_{eq} = L\left(\frac{Q_L{}^2 + 1}{Q_L{}^2}\right)$$

where Q_L is the quality factor of the inductor (L). The **quality factor (Q_L) of the inductor** is calculated from

$$Q_L = \frac{X_L}{R_W}$$

where X_L is the inductive reactance of the inductor (L) at the resonant frequency (f_r), and is calculated from

$$X_L = 2\pi f_r L$$

In most applications, parallel resonant circuits use a **high Q inductor coil** (low R_W resistance). Notice that for a high Q coil ($Q_L > 10$), the parallel equivalent inductance (L_{eq}) is practically equal to the original inductance (L) and the parallel equivalent resistance (R_{eq}) is much larger than the original inductor wire resistance (R_W). Because Q_L is a function of the resonant frequency, the parallel equivalent circuit is only valid at the **resonant frequency (f_r)**.

As in the series resonant circuit, the inductive reactance is equal to the capacitive reactance at the resonant frequency (f_r) in the parallel equivalent circuit. Therefore,

$$X_L = X_C$$

$$2\pi f_r L_{eq} = \frac{1}{2\pi f_r C}$$

$$2\pi f_r L\left(\frac{Q_L^{\ 2}+1}{Q_L^{\ 2}}\right) = \frac{1}{2\pi f_r C}$$

Solving for f_r produces the equation

$$f_r = \frac{1}{2\pi\sqrt{LC}}\sqrt{\frac{Q_L^{\ 2}}{Q_L^{\ 2}+1}}$$

Because most parallel resonant circuits use a high Q coil (Q_L >10), the parallel resonant frequency equation is practically the same as the series resonant frequency equation. Therefore,

$$f_r \cong \frac{1}{2\pi\sqrt{LC}} \qquad \text{(for } Q_L > 10\text{)}$$

When $X_L = X_C$ at the resonant frequency, the inductor current I_L and capacitor current I_C are equal in magnitude and 180 degrees out-of-phase. From Kirchhoff's current law, the total current entering the parallel equivalent L-C network is zero at resonance, and the total current entering the parallel equivalent R-L-C circuit will flow through resistance R_P only. Therefore, the **impedance (Z)** of the parallel resonant circuit at the resonant frequency is equal to resistance R_P, and the phase (θ) between the voltage across the tank circuit (V_a) and the current entering the tank circuit (I) is 0°. Therefore,

$$Z = R_P \qquad \text{(at resonance)}$$

and $$\theta = 0° \qquad \text{(at resonance)}$$

For frequencies below the resonant frequency, the capacitive reactance (X_C) is higher than the inductive reactance (X_L), causing current flow in the parallel L-C network. This will cause the current entering the parallel tank circuit to be higher than at resonance. For frequencies above the resonant frequency, the capacitive reactance (X_C) is lower than the inductive reactance (X_L), causing current flow in the parallel L-C network again. This will cause the current entering the parallel tank circuit to be higher than at resonance again. Therefore, the current entering the parallel tank circuit is at minimum at the resonant frequency and increases at frequencies above and below the resonant frequency, and the impedance is at maximum at the resonant frequency and decreases at frequencies above and below the resonant frequency. This result is the opposite of that observed in the series resonant circuit, which had minimum impedance at the resonant frequency.

The **bandwidth (BW)** of a parallel resonant circuit can be measured from the frequency curve plot by determining the low frequency (f_l) and the high frequency (f_h) at which the impedance of the tank circuit (Z) drops to 0.707 (–3 dB) of the peak value. The bandwidth can be calculated from the equation

$$BW = f_h - f_l$$

The **quality factor (Q)** of the parallel resonant circuit measures how narrow the resonant circuit bandwidth (BW) is in relation to the resonant frequency (f_r). The higher the quality factor, the more narrow the bandwidth in relation to the resonant frequency. The bandwidth (BW) of a parallel resonant circuit can be calculated from the resonant frequency (f_r) and the quality factor (Q) of the tank circuit using the equation

$$BW = \frac{f_r}{Q}$$

If the resistance of the coil is the only resistance in the circuit, then the overall quality factor of the equivalent parallel resonant circuit (Q) will be equal to the quality factor of the inductor (Q_L). In many practical situations, an external load resistance (R) appears in parallel with the tank circuit. If there is a load resistance across the tank circuit, the value of R must be combined with resistance R_{eq} to produce an equivalent parallel resistance (R_P). This will reduce the quality factor of the parallel equivalent resonant circuit. The quality factor of the parallel equivalent resonant circuit can be determined from the ratio of energy stored in the inductor (L) or capacitor (C) to the energy dissipated in the parallel resistance (R_P). This can also be expressed as the ratio of the reactive power to the real power. Therefore,

$$Q = \frac{V_a^2/X_L}{V_a^2/R_P} = \frac{R_P}{X_L}$$

where X_L is calculated from

$$X_L = 2\pi f_r L_{eq} \cong 2\pi f_r L$$

Figure 28-1 Parallel Resonance

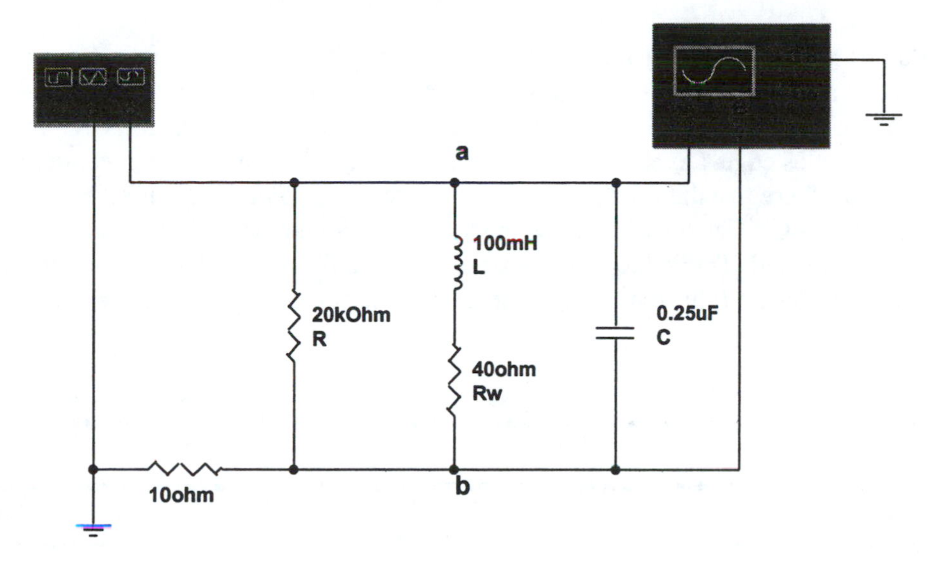

Figure 28-2 Parallel Resonance Frequency Plot

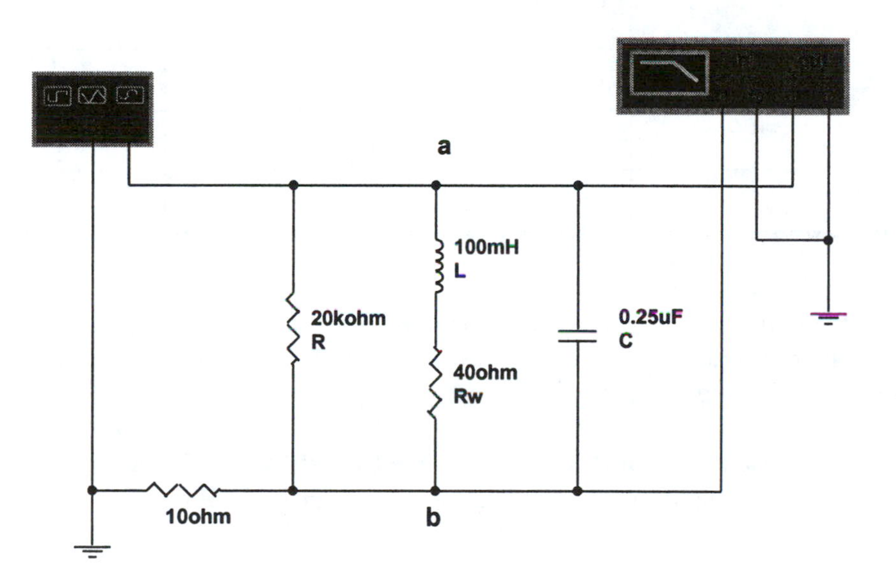

Procedure:

Step 1. Pull down the File menu and open FIG28-1. Bring down the oscilloscope enlargement and make sure that the following settings are selected: Time base (Scale = 200 μs/Div, Xpos = 0, Y/T), Ch A (Scale = 5 V/Div, Ypos = 0, AC), Ch B (Scale = 10 mV/Div, Ypos = 0, AC), Trigger (Pos edge, Level = 1 μV, Nor, A). Bring down the function generator enlargement and make sure that the following settings are selected: *Sine Wave*, Freq = 1 kHz, Ampl = 10 V, Offset = 0. Click the On-Off switch to run the simulation. Click "Pause" when steady state is reached (Tran = 0.044 s or greater). Record the peak voltage V_a at node a (red curve) and V_b at node b (blue curve) in Table 28-1 at the appropriate frequency.

Table 28-1

f (Hz)	V_a (V)	V_b (mV)	I (mA)	Z (kΩ)
800				
900				
950				
1000				
1100				
1200				
1300				

Step 2. Change the frequency of the function generator to each frequency in Table 28-1, run the simulation until steady state is reached, and record each peak value of V_a and V_b for each frequency. Adjust the oscilloscope as needed.

Step 3. Based on each value of V_b in Table 28-1 and the 10 Ω resistor in Figure 28-1, calculate the current (I) at each frequency and record your answers in Table 28-1.

Step 4. Plot the values of the current entering and leaving the tank circuit (I) at each frequency and draw the graph of current (I) as a function of frequency (f).

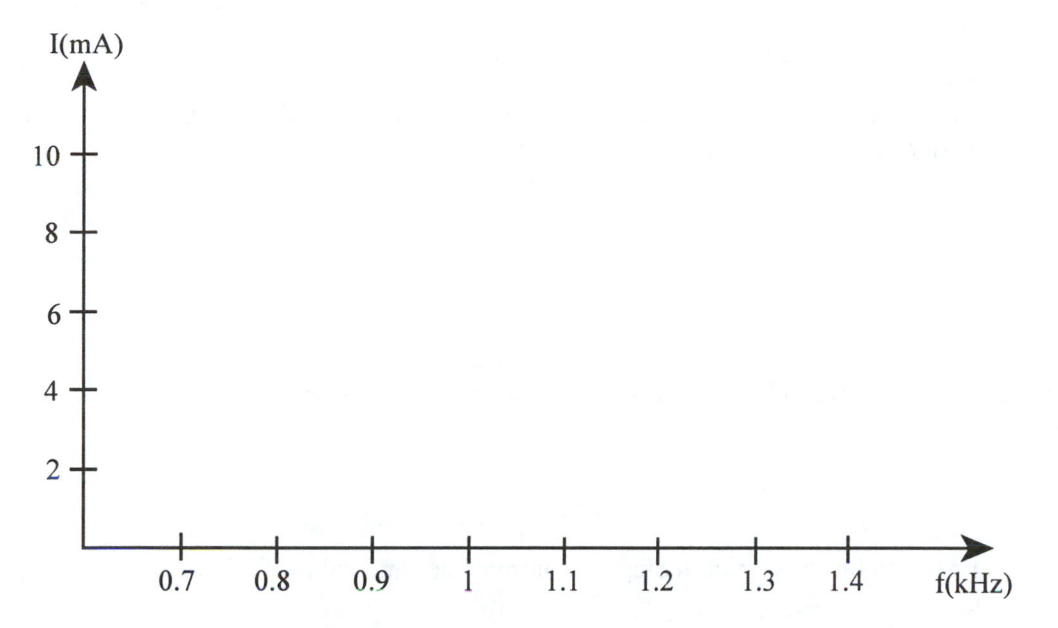

Step 5. Based on each value of V_a and I in Table 28-1, calculate the impedance (Z) of the parallel resonant circuit at each frequency and record your answers in the table.

Step 6. Plot the value of the impedance (Z) at each frequency in Table 28-1 and draw the graph in the space provided.

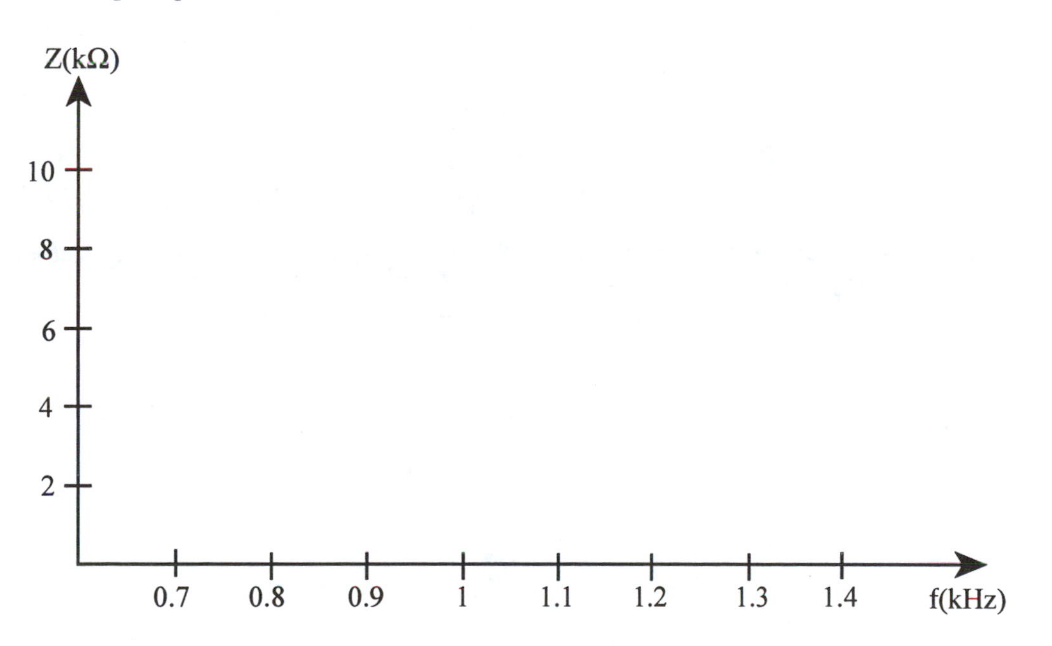

Step 7. Based on the curve plot in Step 6, determine the resonant frequency (f_r) of the parallel resonant circuit.

f_r = ___________

Step 8. Based on the circuit component values in Figure 28-1, calculate the expected resonant frequency (f_r) of the parallel resonant circuit.

Question: How did your calculated resonant frequency in Step 8 compare with the resonant frequency determined from the curve?

Step 9. Based on the curve plot in Step 6, determine the bandwidth (BW) of the parallel resonant circuit.

BW = ___________

Step 10. Based on the circuit component values in Figure 28-1, determine the quality factor (Q_L) of the inductor (L).

Step 11. Based on the circuit component values in Figure 28-1, determine the quality factor (Q) of the parallel resonant circuit.

Question: How did the quality factor (Q_L) of the inductor compare with the quality factor (Q) of the parallel resonant circuit? **Explain any difference.**

Step 12. Based on the quality factor (Q) of the parallel resonant circuit and the resonant frequency (f_r) calculated in Step 8, calculate the expected bandwidth (BW) of this parallel resonant circuit.

Question: How did your calculated bandwidth (BW) in Step 12 compare with the bandwidth measured from the curve in Step 9?

Step 13. Based on the circuit component values in Figure 28-1, calculate the impedance (Z) of the parallel resonant circuit at the resonant frequency.

Questions: How did your calculated impedance at the resonant frequency in Step 13 compare with the impedance at the resonant frequency recorded in Table 28-1?

Based on the graph in Step 6, what conclusion can you draw about the impedance of the parallel resonant circuit as the frequency varies?

Step 14. Change the frequency of the function generator to the resonant frequency (f_r) calculated in Step 8, run the simulation until steady state is reached, and record the phase difference (θ) between the voltage across the tank circuit and current entering and leaving the tank circuit. Change the oscilloscope settings as needed.

θ = ___________

Question: What conclusion can you draw about the phase difference between the voltage and current at resonance? **Explain.**

> *Note:* Steps 15–17 should be performed on the computer using Multisim because a Bode plotter may not be available in a hardwired laboratory.

Step 15. Pull down the File menu and open FIG28-2. Bring down the Bode plotter enlargement and make sure that the following settings are selected: *Magnitude*, Vertical (Lin, F = 1E3, I = 0), Horizontal (Log, F = 1.2 kHz, I = 800 Hz). Click the On-Off switch to run the simulation until the Bode plot is complete, then "pause" the simulation. The Bode plotter is plotting the parallel resonant circuit impedance ($Z = V_a / I$) as a function of frequency. Measure and record the resonant frequency (f_r) and bandwidth (BW) from the curve plot.

f_r = ___________ BW = ___________

Step 16. Change resistance R to 5 kΩ and repeat Step 15. Change the Bode plotter settings as needed.

f_r = ___________ BW = ___________

Question: What conclusion can you draw about the effect of the parallel load resistance (R) on the resonant frequency and the bandwidth? **Explain.**

17. Change resistance R back to 20 kΩ. Change the inductor coil resistance (R_W) to 20 Ω and repeat Step 15. Change the Bode plotter settings as needed.

f_r = ____________ BW = ____________

Question: What conclusion can you draw about the effect of the inductor coil resistance value (R_W) on the resonant frequency and the bandwidth? **Explain.**

Troubleshooting Problems

1. Pull down the file menu and open FIG28-3. Bring down the oscilloscope and click the On-Off switch to run the simulation. Based on the oscilloscope curve plot, which component is open, L or C? **Explain your answer.**

2. Pull down the file menu and open FIG28-4. Bring down the oscilloscope and click the On-Off switch to run the simulation. Pause the simulation when steady state is reached (Tran = 0.044 s). Based on the oscilloscope curve plot, which component is open, L or C? **Explain your answer.**

Name______________________________

Date______________________________

EXPERIMENT

29

High-Pass and Low-Pass Passive Filters

Objectives:

1. Plot the gain and phase response of an R-C low-pass filter.
2. Determine the cutoff frequency of an R-C low-pass filter.
3. Determine how the value of R and C affects the cutoff frequency of an R-C low-pass filter.
4. Plot the gain and phase response of an R-C high-pass filter.
5. Determine the cutoff frequency of an R-C high-pass filter.
6. Determine how the value of R and C affects the cutoff frequency of an R-C high-pass filter.

Materials:

One function generator
One dual-trace oscilloscope
Capacitors — .02 μF, .04 μF
Resistors — 1 kΩ, 2 kΩ

Theory:

A **filter** is a circuit that passes a specific range of frequencies while rejecting other frequencies. A **passive filter** consists of passive circuit elements, such as capacitors, inductors, and resistors. The most common way to describe the frequency response of a filter is to plot the filter voltage gain (V_o/V_{in}) in dB as a function of frequency (f). The frequency at which the output power gain drops to 50% of the maximum value is called the **cutoff frequency (f_c).** When the output power gain drops 50%, the voltage gain drops 3 dB (0.707 of the maximum value). When the filter dB voltage gain is plotted as a function of frequency on a semilog graph using straight lines to approximate the actual frequency response, it is called a **Bode plot**. A Bode plot is an ideal plot of filter frequency response because it assumes that the voltage gain remains constant until the cutoff frequency is reached. The filter network voltage gain in dB is calculated from the actual voltage gain (A) using the equation

$$A_{dB} = 20 \log A$$

where $A = V_o/V_{in}$

There are four basic types of filters, low-pass, high-pass, band-pass, and band-stop. A **low-pass filter** is designed to pass all frequencies below the cutoff frequency and reject all frequencies above the cutoff frequency. A **high-pass filter** is designed to pass all frequencies above the cutoff frequency and reject all frequencies below the cutoff frequency. A **band-pass filter** passes all frequencies within a band of frequencies and rejects all other frequencies outside the band. A **band-stop filter** rejects all frequencies within a band of frequencies and passes all other frequencies outside the band. A band-stop filter is often referred to as a **notch filter**. In this experiment, you will study low-pass and high-pass filters.

A **low-pass R-C filter** is shown in Figure 29-1. At frequencies well below the cutoff frequency, the capacitive reactance of capacitor C is much higher than the resistance of resistor R, causing the output voltage to be practically equal to the input voltage (A = 1) and constant with variations in frequency. At frequencies well above the cutoff frequency, the capacitive reactance of capacitor C is much lower than the resistance of resistor R, causing the output voltage to decrease 20 dB per decade increase in frequency. At the cutoff frequency, the capacitive reactance of capacitor C is equal to the resistance of resistor R, causing the output voltage to be 0.707 times the input voltage (–3 dB). The expected cutoff frequency (f_C) of the low-pass filter in Figure 29-1, based on the circuit component values, can be calculated from

$$X_C = R$$

$$\frac{1}{2\pi f_C C} = R$$

Solving for f_C produces the equation

$$f_C = \frac{1}{2\pi RC}$$

A **high-pass R-C filter** is shown in Figure 29-2. At frequencies well above the cutoff frequency, the capacitive reactance of capacitor C is much lower than the resistance of resistor R, causing the output voltage to be practically equal to the input voltage (A = 1) and constant with variations in frequency. At frequencies well below the cutoff frequency, the capacitive reactance of capacitor C is much higher than the resistance of resistor R, causing the output voltage to decrease 20 dB per decade decrease in frequency. At the cutoff frequency, the capacitive reactance of capacitor C is equal to the resistance of resistor R, causing the output voltage to be 0.707 times the input voltage (–3 dB). The expected cutoff frequency (f_C) of the high-pass filter in Figure 29-2, based on the circuit component values, can also be calculated from

$$f_C = \frac{1}{2\pi RC}$$

When the frequency at the input of a low-pass filter increases above the cutoff frequency, the filter output voltage drops at a constant rate. When the frequency at the input of a high-pass filter decreases below the cutoff frequency, the filter output voltage also drops at a constant rate. The constant drop in filter output voltage per decade increase (×10) or decrease (÷10) in frequency is referred to as **roll-off**. An ideal low-pass or high-pass filter would have an instantaneous drop at the cutoff frequency (f_C), with full signal level on one side of the cutoff frequency and no signal level on the other side of the cutoff frequency. Although the ideal is not achievable, actual filters roll off at –20 dB/decade per pole (R-C circuit). A two-pole filter has two R-C circuits tuned to the same cutoff frequency and rolls off at –40 dB/decade. Each additional pole (R-C circuit) will cause the filter to roll off an additional –20 dB/decade. Therefore, an R-C filter with more poles (R-C circuits) more closely approaches an ideal filter. In a one-pole filter, as shown in Figures 29-1 and 29-2, the phase difference between the input and the output will change by 90 degrees over the frequency range and be 45 degrees at the cutoff frequency. In a two-pole filter, the phase difference will change by 180 degrees over the frequency range and be 90 degrees at the cutoff frequency.

Figure 29-1 Low-Pass R-C Filter

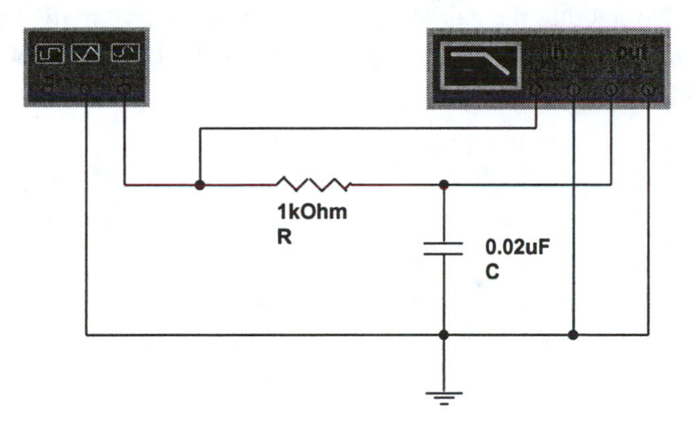

Figure 29-2 High-Pass R-C Filter

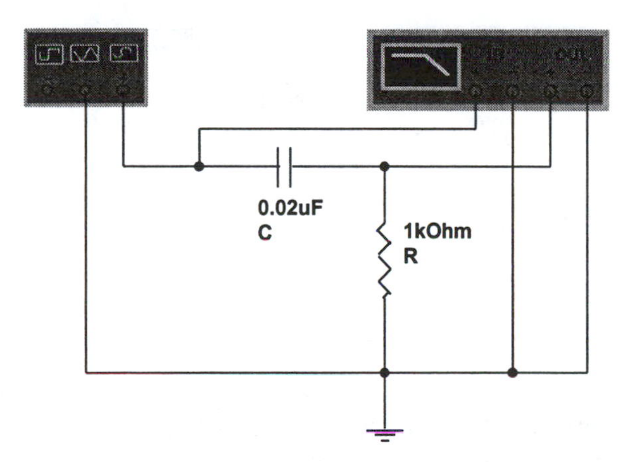

Procedure:

Low-Pass Filter

Step 1. Pull down the File menu and open FIG29-1. Bring down the Bode plotter enlargement and make sure that the following settings are selected: *Magnitude*, Vertical (Log, F = 0 dB, I = –40 dB), Horizontal (Log, F = 1 MHz, I = 100 Hz). You will plot the gain-frequency response in dB for the R-C low-pass filter as the frequency varies between 100 Hz and 1 MHz using the Bode plotter.

> *Note:* If you are performing this experiment in a laboratory environment, you may not have a Bode plotter available. You will need to plot the frequency response curves by making measurements at different frequencies using an oscilloscope and plotting the curves on semilog graph paper.

Step 2. Click the On-Off switch to run the simulation. "Pause" the simulation after the Bode plot is complete. Notice that the gain-frequency response curve in dB has been plotted between the frequencies of 100 Hz and 1 MHz by the Bode plotter. Sketch the curve plot in the space provided.

A_{dB}

f

Question: Is the frequency response curve plotted in Step 2 that of a low-pass filter? **Explain.**

Step 3. Move the cursor to a flat part of the curve at a frequency of approximately 100 Hz. Record the voltage gain in dB on the curve plot.

Step 4. Calculate the actual voltage gain (A) from the dB voltage gain (A_{dB}) measured in Step 3.

Question: Was the voltage gain on the flat part of the gain-frequency response curve what you expected for the circuit in Figure 29-1? **Explain why.**

Step 5. Move the cursor as close as possible to a point on the curve that is 3 dB down from the dB gain at 100 Hz. Record the frequency (cutoff frequency, f_C) on the curve plot.

Step 6. Calculate the expected cutoff frequency (f_C) based on the circuit component values in Figure 29-1.

Question: How did the calculated value for the cutoff frequency compare with the measured value recorded on the curve plot in Step 5?

Step 7. Move the cursor to a point on the curve that is as close as possible to ten times f_C. Record the dB gain and frequency (f_2) on the curve plot in Step 2.

Question: How much did the dB gain decrease for a one-decade increase in frequency? Was it what you expected for a single pole (single R-C) low-pass filter?

Step 8. Click "Phase" on the Bode plotter to plot the phase curve. Make sure that the vertical axis initial value (I) is –90° and the final value (F) is 0°. Click the On-Off switch to run the simulation again. You are looking at the phase difference (θ) between the filter input and output as a function of frequency. Sketch the curve plot in the space provided.

θ

f

Step 9. Move the cursor to approximately 100 Hz and 1 MHz and record the phase (θ) in degrees on the curve plot for each frequency. Next, move the cursor as close as possible on the curve to the cutoff frequency (f_C). Record the frequency and phase on the curve plot.

Questions: Was the phase at the cutoff frequency (f_C) what you expected for a single pole (single R-C) low-pass filter?

Did the phase change with frequency? Is this expected for an R-C low-pass filter?

Step 10. Change the value of resistor R to 2 kΩ in Figure 29-1. Click "Magnitude" on the Bode plotter. Click the On-Off switch to run the simulation. "Pause" the simulation after the Bode plot is complete. Measure the cutoff frequency (f_C) following the procedure in Step 5 and record your answer.

f_C = _______________

Step 11. Based on the new value of resistor R, calculate the new cutoff frequency (f_C).

Questions: How did the calculated value of the new cutoff frequency compare with the value measured in Step 10?

What effect did changing the value of resistor R have on the cutoff frequency? **Explain.**

Step 12. Change the value of capacitor C to 0.04 μF in Figure 29-1. Click the On-Off switch to run the simulation. Measure the cutoff frequency (f_C) following the procedure in Step 5 and record your answer.

f_C = _______________

Step 13. Based on the new value of resistor R and capacitor C, calculate the new cutoff frequency (f_C).

Questions: How did the calculated value of the new cutoff frequency compare with the value measured in Step 12?

What effect did changing the value of capacitor C have on the cutoff frequency? **Explain.**

High-Pass Filter

Step 14. Pull down the File menu and open FIG29-2. Bring down the Bode plotter enlargement and make sure that the following settings are selected: *Magnitude*, Vertical (Log, F = 0 dB, I = –40 dB), Horizontal (Log, F = 1 MHz, I = 100 Hz). You will plot the gain-frequency response in dB for an R-C high-pass filter as the frequency varies between 100 Hz and 1 MHz using the Bode plotter.

Note: If you are performing this experiment in a laboratory environment, you may not have a Bode plotter available. You will need to plot the frequency response curves by making measurements at different frequencies using an oscilloscope and plotting the curves on semilog graph paper.

Step 15. Click the On-Off switch to run the simulation. "Pause" the simulation after the Bode plot is complete. Notice that the gain-frequency response curve in dB has been plotted between the frequencies of 100 Hz and 1 MHz by the Bode plotter. Sketch the curve plot in the space provided.

A_{dB}

f

Question: Is the frequency response curve plotted in Step 15 that of a high-pass filter? **Explain.**

Step 16. Move the cursor to a flat part of the curve at a frequency of approximately 1 MHz. Record the voltage gain in dB on the curve plot.

Step 17. Calculate the actual voltage gain (A) from the dB voltage gain (A_{dB}) measured in Step 16.

Question: Was the voltage gain on the flat part of the gain-frequency response curve what you expected for the circuit in Figure 29-2? **Explain why.**

Step 18. Move the cursor as close as possible to a point on the curve that is 3 dB down from the dB gain at 1 MHz. Record the frequency (cutoff frequency, f_C) on the curve plot.

Step 19. Calculate the expected cutoff frequency (f_C) based on the circuit component values in Figure 29-2.

Question: How did the calculated value of the cutoff frequency compare with the measured value recorded on the curve plot in Step 18?

Step 20. Move the cursor to a point on the curve that is as close as possible to one-tenth f_C. Record the dB gain and frequency (f_2) on the curve plot in Step 15.

Question: How much did the dB gain decrease for a one-decade decrease in frequency? Was it what you expected for a single pole (single R-C) high-pass filter?

Step 21. Click "Phase" on the Bode plotter to plot the phase curve. Make sure that the vertical axis initial value (I) is 0° and the final value (F) is 90°. Click the On-Off switch to run the simulation again. You are looking at the phase difference (θ) between the filter input and output as a function of frequency. Sketch the curve plot in the space provided.

θ

f

Step 22. Move the cursor to approximately 100 Hz and 1 MHz and record the phase in degrees on the curve plot for each frequency. Next, move the cursor as close as possible on the curve to the cutoff frequency (f_C). Record the frequency and phase on the curve plot.

Questions: Was the phase at the cutoff frequency (f_C) what you expected for a single pole (single R-C) high-pass filter?

Did the phase change with frequency? Is this expected for an R-C high-pass filter?

Step 23. Change the value of resistor R to 2 kΩ in Figure 29-2. Click "Magnitude" on the Bode plotter. Click the On-Off switch to run the simulation. Measure the cutoff frequency (f_C) following the procedure in Step 18 and record your answer.

f_C = _______________

Step 24. Based on the new value of resistor R, calculate the new cutoff frequency (f_C).

Questions: How did the calculated value of the new cutoff frequency compare with the value measured in Step 23?

What effect did changing the value of resistor R have on the cutoff frequency? **Explain.**

Step 25. Change the value of capacitor C to 0.04 μF in Figure 29-2. Click the On-Off switch to run the simulation. Measure the cutoff frequency (f_C) following the procedure in Step 18 and record your answer.

f_C = _______________

Step 26. Based on the new value of resistor R and capacitor C, calculate the new cutoff frequency (f_C).

Questions: How did the calculated value of the new cutoff frequency compare with the value measured in Step 25?

What effect did changing the value of capacitor C have on the cutoff frequency? **Explain.**

Troubleshooting Problems

1. Pull down the File menu and open FIG29-3. Click the On-Off switch to run the simulation. Based on the curve plots on the oscilloscope and the Bode plotter, determine the defective component and the defect (short or open).

   Defective component ____________ Defect ___________

2. Pull down the File menu and open FIG29-4. Click the On-Off switch to run the simulation. Based on the curve plots on the oscilloscope and the Bode plotter, determine the defective component and the defect (short or open).

   Defective component ____________ Defect ___________

Name________________________________

Date________________________________

EXPERIMENT

30

Band-Pass and Band-Stop Passive Filters

Objectives:

1. Plot the gain frequency response of an L-C series resonant and an L-C parallel resonant band-pass filter.
2. Determine the center frequency and the bandwidth of the L-C band-pass filters.
3. Determine how the circuit resistance affects the bandwidth of an L-C band-pass filter.
4. Plot the gain frequency response of an L-C series resonant and an L-C parallel resonant band-stop (notch) filter.
5. Determine the center frequency and the bandwidth of the L-C band-stop filters.
6. Determine how the circuit resistance affects the bandwidth of an L-C band-stop filter.

Materials:

One function generator
One dual-trace oscilloscope
One 0.25 μF capacitor
One 100 mH inductor
Resistors—100 Ω, 4 kΩ, 5 kΩ , 200 kΩ

Theory:

You should complete Experiments 27 and 28 on resonance before attempting this experiment on band-pass and band-stop filters. If you have completed those experiments, review their Theory sections.

A **filter** is a circuit that passes a specific range of frequencies while rejecting other frequencies. A **passive filter** consists of passive circuit elements, such as capacitors, inductors, and resistors. The most common way to describe the frequency response of a filter is to plot the filter voltage gain (V_o/V_{in}) in dB as a function of frequency (f). The frequency at which the output power gain drops to 50% of the maximum value is called the **cutoff frequency (f_c).** When the output power gain drops 50%, the voltage gain drops 3 dB (0.707 of the maximum value). When the filter dB voltage gain is plotted as a function of frequency on a semilog graph using straight lines to approximate the actual frequency response, it is called a **Bode plot**. A Bode plot is an ideal plot of filter frequency response because it assumes that the voltage gain remains constant until the cutoff frequency is reached. The filter network voltage gain in dB is calculated from the actual voltage gain (A) using the equation

$$A_{dB} = 20 \log A$$

where $A = V_o/V_{in}$

There are four basic types of filters, low-pass, high-pass, band-pass, and band-stop. A **low-pass filter** is designed to pass all frequencies below the cutoff frequency and reject all frequencies above the cutoff frequency. A **high-pass filter** is designed to pass all frequencies above the cutoff frequency and reject all frequencies below the cutoff frequency. A **band-pass filter** passes all frequencies within a band of frequencies and rejects all other frequencies outside the band. A **band-stop filter** rejects all frequencies within a band of frequencies and passes all other frequencies outside the band. A band-stop filter is often referred to as a **notch filter**. In this experiment, you will study band-pass and band-stop (notch) filters.

An **L-C series resonant band-pass filter** is shown in Figure 30-1. The impedance of the series L-C circuit is lowest at the resonant frequency and increases on both sides of the resonant frequency. (See Experiment 27.) This will cause the output voltage (V_o) to be highest at the resonant frequency and decrease on both sides of the resonant frequency. An **L-C parallel resonant band-pass filter** is shown in Figure 30-2. The impedance of the parallel L-C circuit is highest at the resonant frequency and decreases on both sides of the resonant frequency. (See Experiment 28.) This will cause the output voltage (V_o) to be highest at the resonant frequency and decrease on both sides of the resonant frequency

An **L-C series resonant band-stop (notch) filter** is shown in Figure 30-3. The impedance of the series L-C circuit is lowest at the resonant frequency and increases on both sides of the resonant frequency. This will cause the output voltage (V_o) to be lowest at the resonant frequency and increase on both sides of the resonant frequency. An **L-C parallel resonant band-stop (notch) filter** is shown in Figure 30-4. The impedance of the parallel L-C circuit is highest at the resonant frequency and decreases on both sides of the resonant frequency. This will cause the output voltage (V_o) to be lowest at the resonant frequency and increase on both sides of the resonant frequency.

The **center frequency (f_O)** for the L-C series resonant and the L-C parallel resonant band-pass and band-stop (notch) filters is equal to the resonant frequency of the L-C circuit, which can be calculated from

$$f_O = \frac{1}{2\pi\sqrt{LC}}$$

For the L-C parallel resonant filter, the equation is accurate only for a high Q inductor coil ($Q_L \geq 10$), where Q_L is calculated from

$$Q_L = \frac{X_L}{R_W}$$

and X_L is the inductive reactance at the resonant frequency (center frequency, f_O) and R_W is the inductor coil resistance. (See Experiment 28.)

In the band-pass and band-stop (notch) filters, the low cutoff frequency (f_{C1}) and the high cutoff frequency (f_{C2}) on the gain-frequency plot are the frequencies where the voltage gain has dropped 3 dB (0.707) from the highest dB gain. The filter bandwidth (BW) is the difference between the high cutoff frequency (f_{C2}) and the low cutoff frequency (f_{C1}). Therefore,

$$BW = f_{C2} - f_{C1}$$

The center frequency (f_O) is the geometric mean of the low cutoff frequency and the high cutoff frequency. Therefore,

$$f_O = \sqrt{f_{C1} f_{C2}}$$

The quality factor (Q) of the band-pass and band-stop (notch) filters is the ratio of the center frequency (f_O) and the bandwidth (BW), and it is an indication of the selectivity of the filter. Therefore,

$$Q = \frac{f_O}{BW}$$

A higher value of Q means a narrower bandwidth and a more selective filter.

The quality factor (Q) of a series resonant filter is determined by first calculating the inductive reactance (X_L) of the inductor at the resonant frequency (center frequency, f_O), and then dividing the inductive reactance by the total series resistance (R_T). Therefore,

$$Q = \frac{X_L}{R_T}$$

where $X_L = 2\pi f_O L$

The quality factor (Q) of a parallel resonant filter is determined by first calculating the inductive reactance (X_L) of the inductor at the resonant frequency (center frequency, f_O), and then dividing the total parallel resistance (R_P) by the inductive reactance (X_L). Therefore,

$$Q = \frac{R_P}{X_L}$$

Because the inductor wire resistance (R_W) is in series with inductor L, the circuits in Figures 30-2 and 30-4 are not exactly parallel resonant circuits. In order to make them be true parallel resonant circuits, the series combination of inductance (L) and resistance (R_W) must first be converted into an equivalent parallel network with resistance R_{eq} in parallel with inductance L. In Figure 30-2, the parallel equivalent resistance (R_{eq}) will also be in parallel with resistor R and resistor R_S, making the total resistance of the parallel resonant circuit (R_P) equal to the parallel equivalent of resistors R, R_S, and R_{eq}. Therefore, R_P can be solved from

$$\frac{1}{R_P} = \frac{1}{R} + \frac{1}{R_S} + \frac{1}{R_{eq}}$$

In Figure 30-4, the parallel equivalent resistance (R_{eq}) will be in parallel with resistor R, making the total resistance of the parallel resonant circuit (R_P) equal to the parallel equivalent of resistors R and R_{eq}. Therefore, R_P can be solved from

$$R_P = \frac{RR_{eq}}{R + R_{eq}}$$

The equation for converting resistance R_W to equivalent parallel resistance R_{eq} is

$$R_{eq} = R_W(Q_L{}^2 + 1)$$

Figure 30-1 L-C Series Resonant Band-Pass Filter

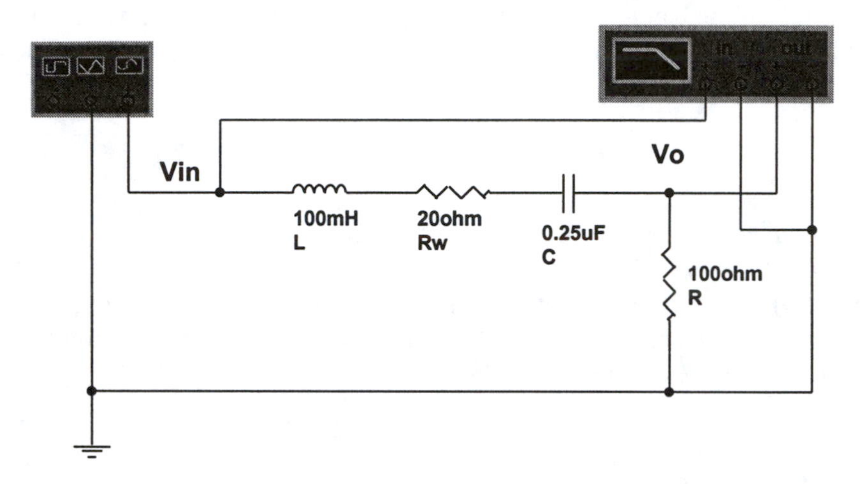

Figure 30-2 L-C Parallel Resonant Band-Pass Filter

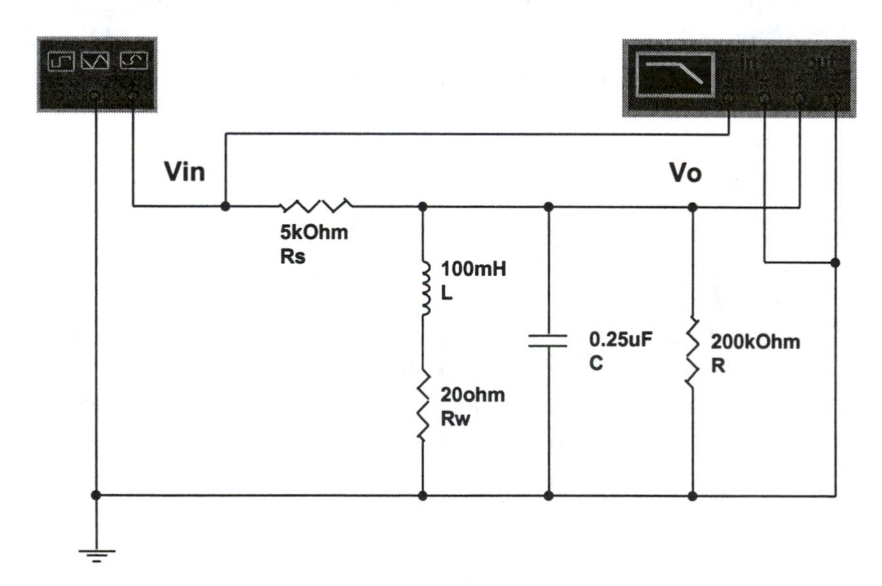

Figure 30-3 L-C Series Resonant Band-Stop (Notch) Filter

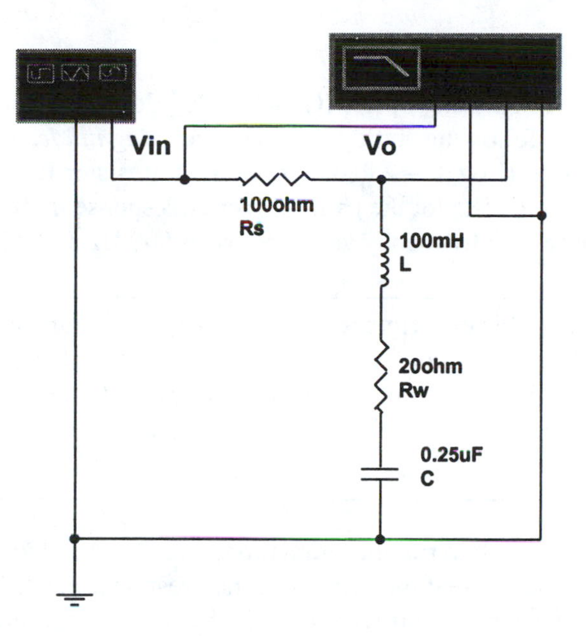

Figure 30-4 L-C Parallel Resonant Band-Stop (Notch) Filter

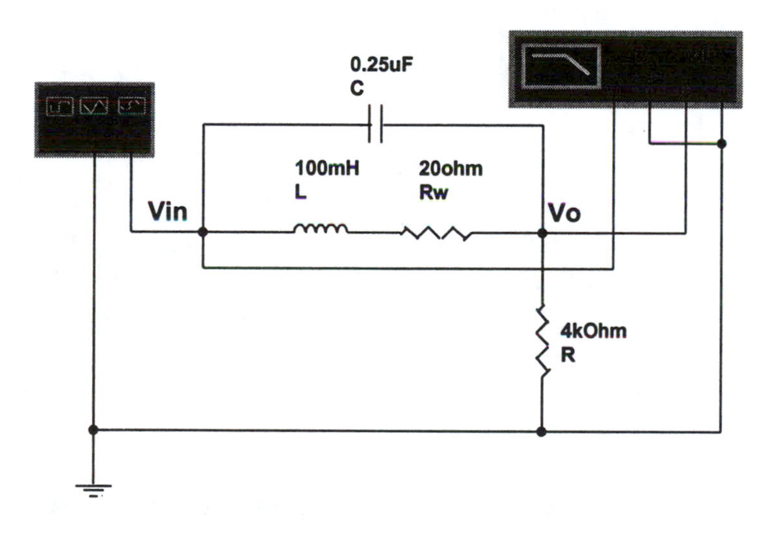

Procedure:

Band-Pass Filters

Step 1. Pull down the File menu and open FIG30-1. Bring down the Bode plotter enlargement and make sure that the following settings are selected: *Magnitude*, Vertical (Log, F = 0 dB, I = –20 dB), Horizontal (Log, F = 2 kHz, I = 500 Hz). Resistor R_W is simulating the inductor wire resistance. You will plot the gain-frequency response in dB for an **L-C series resonant band-pass filter** as the frequency varies between 500 Hz and 2 kHz using the Bode plotter.

> *Note:* If you are performing this experiment in a laboratory environment, you may not have a Bode plotter available. You will need to plot the frequency response curves by making measurements at different frequencies using an oscilloscope and plotting the curves on semilog graph paper. Replace resistance R_W in Figure 30-1 with the actual inductor wire resistance.

Step 2. Click the On-Off switch to run the simulation, then "pause" the simulation after the Bode plot is complete. Notice that the gain-frequency response curve in dB has been plotted between the frequencies of 500 Hz and 2 kHz by the Bode plotter. Sketch the curve plot in the space provided.

A_{dB}

f

Question: Is the frequency response curve plotted in Step 2 that of a band-pass filter? **Explain why.**

Step 3. Move the cursor to the center of the curve at its peak point. Record the center frequency (f_o) and the voltage gain in dB on the curve plot.

Step 4. Based on the dB voltage gain measured in Step 3, calculate the actual voltage gain (A) of the series resonant band-pass filter at the center frequency.

Step 5. Move the cursor as close as possible to a point on the left side of the curve that is 3 dB down from the dB gain measured at frequency f_O. Record the approximate frequency (low cutoff frequency, f_{C1}) on the curve plot. Next, move the cursor as close as possible to a point on the right side of the curve that is 3 dB down from the dB gain measured at frequency f_O. Record the approximate frequency (high cutoff frequency, f_{C2}) on the curve plot.

Step 6. Based on the values of f_{C1} and f_{C2} measured on the curve plot in Step 2, determine the bandwidth (BW) of the series resonant band-pass filter and record your answer on the curve plot.

Step 7. Based on the circuit component values in Figure 30-1, calculate the expected center frequency (f_O) of the series resonant band-pass filter.

Question: How did the calculated value of the center frequency (f_O) based on the circuit component values compare with the measured value recorded on the curve plot in Step 2?

Step 8. Based on the values of f_{C1} and f_{C2}, calculate the center frequency (f_O).

Question: How did the calculated value of the center frequency (f_O) based on f_{C1} and f_{C2} compare with the measured value recorded on the curve plot in Step 2?

Step 9. Based on the circuit component values, calculate the quality factor (Q) of the series resonant band-pass filter.

Step 10. Based on the circuit quality factor (Q) and the center frequency (f_o), calculate the expected bandwidth (BW) of the series resonant band-pass filter.

Question: How did the expected bandwidth calculated from the value of Q and the center frequency compare with the bandwidth measured on the curve plot in Step 2?

Step 11. Change resistance R to 200 Ω. Click the On-Off switch to run the simulation again. Measure the center frequency (f_o) and the bandwidth (BW) from the curve plot and record the values.

f_o = ______________ BW = ______________

Step 12. Based on the new value of R, calculate the new circuit quality factor (Q).

Step 13. Based on the new circuit quality factor (Q), calculate the new bandwidth (BW).

Questions: How did the new bandwidth calculated in Step 13 compare with the bandwidth measured in Step 11?

What effect did changing the resistance of R have on the center frequency of the series resonant band-pass filter?

What effect did changing the resistance of R have on the bandwidth of the series resonant band-pass filter? **Explain why.**

Step 14. Pull down the File menu and open FIG30-2. Bring down the Bode plotter enlargement and make sure that the following settings are selected: *Magnitude*, Vertical (Log, F = 0 dB, I = –20 dB), Horizontal (Log, F = 2 kHz, I = 500 Hz). Resistor R_W is simulating the inductor wire resistance. You will plot the gain-frequency response in dB for an **L-C parallel resonant band-pass filter** as the frequency varies between 500 Hz and 2 kHz using the Bode plotter.

> *Note:* If you are performing this experiment in a laboratory environment, you may not have a Bode plotter available. You will need to plot the frequency response curves by making measurements at different frequencies using an oscilloscope and plotting the curves on semilog graph paper. Replace resistance R_W in Figure 30-2 with the actual inductor wire resistance.

Step 15. Click the On-Off switch to run the simulation, then "pause" the simulation after the Bode plot is complete. Notice that the gain-frequency response curve in dB has been plotted between the frequencies of 500 Hz and 2 kHz by the Bode plotter. Sketch the curve plot in the space provided.

A_{dB}

f

Question: Is the frequency response curve plotted in Step 15 that of a band-pass filter? **Explain why.**

Step 16. Move the cursor to the center of the curve at its peak point. Record the center frequency (f_O) and the voltage gain in dB on the curve plot.

Step 17. Based on the dB voltage gain measured in Step 16, calculate the actual voltage gain (A) of the parallel resonant band-pass filter at the center frequency.

Step 18. Move the cursor as close as possible to a point on the left side of the curve that is 3 dB down from the dB gain measured at frequency f_O. Record the approximate frequency (low cutoff frequency, f_{C1}) on the curve plot. Next, move the cursor as close as possible to a point on the right side of the curve that is 3 dB down from the dB gain measured at frequency f_O. Record the approximate frequency (high cutoff frequency, f_{C2}) on the curve plot.

Step 19. Based on the values of f_{C1} and f_{C2} measured in Step 18, determine the bandwidth (BW) of the parallel resonant band-pass filter and record your answer on the curve plot.

Step 20. Based on the circuit component values in Figure 30-2, calculate the expected center frequency (f_O) of the parallel resonant band-pass filter.

Question: How did the calculated value of the center frequency (f_O) based on the circuit component values compare with the measured value recorded on the curve plot in Step 15?

Step 21. Based on the values of f_{C1} and f_{C2}, calculate the center frequency (f_O).

Question: How did the calculated value of the center frequency (f_O) based on f_{C1} and f_{C2} compare with the measured value recorded on the curve plot in Step 15?

Step 22. Based on the value of L and R_W, calculate the quality factor (Q_L) of the inductor.

Step 23. Based on the quality factor (Q_L) of the inductor and R_W, calculate the equivalent parallel inductor resistance (R_{eq}) across the tank circuit.

Step 24. Based on the value of R_{eq}, R_S, and R, calculate the total parallel resistance (R_P) across the tank circuit.

Step 25. Based on the value of R_P, calculate the quality factor (Q) of the parallel resonant band-pass filter.

Step 26. Based on the filter quality factor (Q) and the center frequency (f_O), calculate the expected bandwidth (BW) of the parallel resonant band-pass filter.

Question: How did the expected bandwidth calculated from the value of Q and the center frequency compare with the bandwidth measured on the curve plot in Step 15?

Step 27. Change resistance R to 5 kΩ. Click the On-Off switch to run the simulation again. Measure the center frequency (f_O) and the bandwidth (BW) from the curve plot and record the values.

f_O = ______________ BW = ______________

Questions: What effect did changing the resistance of R have on the center frequency of the parallel resonant band-pass filter?

What effect did changing the resistance of R have on the bandwidth of the parallel resonant band-pass filter? **Explain why.**

Band-Stop (Notch) Filters

Step 28. Pull down the File menu and open FIG30-3. Bring down the Bode plotter enlargement and make sure that the following settings are selected: *Magnitude*, Vertical (Log, F = 0 dB, I = –20 dB), Horizontal (Log, F = 2 kHz, I = 500 Hz). Resistor R_W is simulating the inductor wire resistance. You will plot the gain-frequency response in dB for an **L-C series resonant band-stop (notch) filter** as the frequency varies between 500 Hz and 2 kHz using the Bode plotter.

Note: If you are performing this experiment in a laboratory environment, you may not have a Bode plotter available. You will need to plot the frequency response curves by making measurements at different frequencies using an oscilloscope and plotting the curves on semilog graph paper. Replace resistance R_W in Figure 30-3 with the actual inductor wire resistance.

Step 29. Click the On-Off switch to run the simulation, then "pause" the simulation after the Bode plot is complete. Notice that the gain-frequency response curve in dB has been plotted between the frequencies of 500 Hz and 2 kHz by the Bode plotter. Sketch the curve plot in the space provided.

A_{dB}

f

Question: Is the frequency response curve plotted in Step 29 that of a band-stop (notch) filter? **Explain why**.

Step 30. Move the cursor to the center of the curve at its lowest point. Record the center frequency (f_O) and the voltage gain in dB on the curve plot.

Step 31. Based on the dB voltage gain measured in Step 30, calculate the actual voltage gain (A) of the filter at the center frequency.

Question: How did the voltage gain (A) at the center frequency of the series resonant band-stop (notch) filter compare with the voltage gain (A) at the center frequency of the series resonant band-pass filter in Step 4? Was this expected?

Step 32. Move the cursor as close as possible to a point on the left side of the curve that is 3 dB down from the highest dB gain on the flat part of the curve. Record the approximate frequency (low cutoff frequency, f_{C1}) on the curve plot. Next, move the cursor as close as possible to a point on the right side of the curve that is 3 dB down from the highest dB gain on the flat part of the curve. Record the approximate frequency (high cutoff frequency, f_{C2}) on the curve plot.

Step 33. Based on the values of f_{C1} and f_{C2} measured in Step 32, determine the bandwidth (BW) of the series resonant band-stop (notch) filter and record your answer on the curve plot.

Step 34. Based on the circuit component values in Figure 30-3, calculate the expected center frequency (f_O) of the series resonant band-stop (notch) filter.

Question: How did the calculated value of the center frequency (f_O) based on the circuit component values compare with the measured value recorded on the curve plot in Step 29?

Step 35. Based on the values of f_{C1} and f_{C2}, calculate the center frequency (f_O).

Question: How did the calculated value of the center frequency (f_O) based on f_{C1} and f_{C2} compare with the measured value recorded on the curve plot in Step 29?

Step 36. Based on the circuit component values, calculate the quality factor (Q) of the series resonant band-stop (notch) filter.

Step 37. Based on the circuit quality factor (Q) and the center frequency (f_O), calculate the expected bandwidth (BW) of the series resonant band-stop (notch) filter.

Question: How did the expected bandwidth calculated from the value of Q and the center frequency compare with the bandwidth measured on the curve plot in Step 29?

Step 38. Pull down the File menu and open FIG30-4. Bring down the Bode plotter enlargement and make sure that the following settings are selected: *Magnitude*, Vertical (Log, F = 0 dB, I = –20 dB), Horizontal (Log, F = 2 kHz, I = 500 Hz). Resistor R_W is simulating the inductor wire resistance. You will plot the gain-frequency response in dB for an **L-C parallel resonant band-stop (notch) filter** as the frequency varies between 500 Hz and 2 kHz using the Bode plotter.

> *Note:* If you are performing this experiment in a laboratory environment, you may not have a Bode plotter available. You will need to plot the frequency response curves by making measurements at different frequencies using an oscilloscope and plotting the curves on semilog graph paper. Replace resistance R_W in Figure 30-4 with the actual inductor wire resistance.

Step 39. Click the On-Off switch to run the simulation, then "pause" the simulation after the Bode plot is complete. Notice that the gain-frequency response curve in dB has been plotted between the frequencies of 500 Hz and 2 kHz by the Bode plotter. Sketch the curve plot in the space provided.

A_{dB}

f

Question: Is the frequency response curve plotted in Step 39 that of a band-stop (notch) filter? **Explain why**.

Step 40. Move the cursor to the center of the curve at its lowest point. Record the center frequency (f_O) and the voltage gain in dB on the curve plot.

Step 41. Based on the dB voltage gain measured in Step 40, calculate the actual voltage gain (A) of the filter at the center frequency.

Question: How did the voltage gain (A) at the center frequency of the parallel resonant band-stop (notch) filter compare with the voltage gain (A) at the center frequency of the parallel resonant band-pass filter in Step 17? **Explain**.

Step 42. Move the cursor as close as possible to a point on the left side of the curve that is 3 dB down from the highest dB gain on the flat part of the curve. Record the approximate frequency (low cutoff frequency, f_{C1}) on the curve plot. Next, move the cursor as close as possible to a point on the right side of the curve that is 3 dB down from the highest dB gain on the flat part of the curve. Record the approximate frequency (high cutoff frequency, f_{C2}) on the curve plot.

Step 43. Based on the values of f_{C1} and f_{C2} measured in Step 42, determine the bandwidth (BW) of the parallel resonant band-stop (notch) filter and record your answer on the curve plot.

Step 44. Based on the circuit component values in Figure 30-4, calculate the expected center frequency (f_O) of the parallel resonant band-stop (notch) filter.

Question: How did the calculated value of the center frequency (f_O) based on the circuit component values compare with the measured value recorded on the curve plot in Step 39?

Step 45. Based on the values of f_{C1} and f_{C2}, calculate the center frequency (f_O).

Question: How did the calculated value of the center frequency (f_O) based on f_{C1} and f_{C2} compare with the measured value recorded on the curve plot in Step 39?

Step 46. Based on the value of L and R_W, calculate the quality factor (Q_L) of the inductor.

Step 47. Based on the quality factor (Q_L) of the inductor, calculate the equivalent parallel inductor resistance (R_{eq}) across the tank circuit.

Step 48. Based on the value of R_{eq} and R, calculate the total parallel resistance (R_P) across the tank circuit.

Step 49. Based on the value of R_P, calculate the quality factor (Q) of the parallel resonant band-stop (notch) filter.

Step 50. Based on the filter quality factor (Q) and the center frequency (f_O), calculate the expected bandwidth (BW) of the parallel resonant band-stop (notch) filter.

Question: How did the expected bandwidth calculated from the value of Q and the center frequency compare with the bandwidth measured on the curve plot in Step 39?

Step 51. Change resistance R to 2 kΩ. Click the On-Off switch to run the simulation again. Measure the center frequency (f_O) and the bandwidth (BW) from the curve plot and record the values.

f_O = ______________ BW = ______________

Questions: What effect did changing the resistance of R have on the center frequency of the parallel resonant band-stop (notch) filter?

What effect did changing the resistance of R have on the bandwidth of the parallel resonant band-stop (notch) filter? **Explain why**.

Troubleshooting Problems

1. Pull down the File menu and open FIG30-5. Click the On-Off switch to run the simulation. Based on the curve plot on the Bode plotter, determine the defective component and the defect (short or open).

   Defective component ___________ Defect ___________

2. Pull down the File menu and open FIG30-6. Click the On-Off switch to run the simulation. Based on the curve plot on the Bode plotter, determine the defective component and the defect (short or open).

   Defective component ___________ Defect ___________

3. Pull down the File menu and open FIG30-7. Click the On-Off switch to run the simulation. Based on the curve plot on the Bode plotter, determine the defective component and the defect (short or open).

   Defective component ___________ Defect __________

4. Pull down the File menu and open FIG30-8. Click the On-Off switch to run the simulation. Based on the curve plot on the Bode plotter, determine the defective component and the defect (short or open).

   Defective component ___________ Defect __________

Appendix

Notes on Using Electronics Workbench Multisim

1. If you wish to remove a component from a circuit, disconnect both terminals from the circuit; otherwise, you may get an error message.

2. If you wish to pause a simulation on the oscilloscope screen, click the pause symbol (11) next to the On-Off switch. To resume the paused simulation, click the pause symbol (11) again.

3. If you wish to measure the resistance of a component using the multimeter, remove the component from the circuit and connect the multimeter across the component terminals. Make sure you connect a ground symbol to one of the component terminals.

4. The circuit disk provided with this manual is write protected; therefore, you cannot save a changed circuit to the disk. If you wish to save a changed circuit, you must select *Save as* in the File menu and save it on the hard drive or another disk.

5. The color of an oscilloscope curve trace is the same color as the circuit wire connected to the oscilloscope input. A wire color can be changed by right clicking the wire with the mouse, selecting 'color,' and selecting a new color from the option table on the screen.

6. The speed of a simulation can be changed by selecting "Default Instrument Settings" in the Simulate menu and changing the Maximum time step.

7. You can change a component value by double-clicking it with the mouse and changing the menu value using the keyboard.

8. You can bring down an instrument enlargement by double-clicking the instrument with the mouse.

9. Batteries and switches do not have resistance. To convert an ideal battery (no internal resistance) to a real battery (has internal resistance), add a small resistor in series with the battery.

10. The Multisim Help menu has all the information needed to get started using Multisim.

Bibliography

Bell, D. A. *Electronic Circuits: Principles, Applications, and Computer Analysis*. Upper Saddle River, NJ: Prentice Hall, 1995.

Boylestad, R. L. *Introductory Circuit Analysis*. 10th ed. Upper Saddle River, NJ: Prentice Hall, 2003.

Cook, N. P. *Introductory DC/AC Circuits*. 5th ed. Upper Saddle River, NJ: Prentice Hall, 2002.

Floyd, T. L. *Electric Circuits Fundamentals*. 5th ed. Upper Saddle River, NJ: Prentice Hall, 2001.

Floyd, T. L. *Principles of Electric Circuits*. 7th ed. Upper Saddle River, NJ: Prentice Hall, 2003.

Paynter, R.T. *Introductory Electric Circuits*. Upper Saddle River, NJ: Prentice Hall, 1999.

Robbins, A. H., and Miller, W. C. *Circuit Analysis: Theory and Practice*. 2nd ed. Albany, NY: Delmar, 2000.